M.J. HARRIS

Statistical Mechanics
of Phase Transitions

Statistical Mechanics
of Phase Transitions

J. M. Yeomans

*Department of Physics,
University of Oxford*

CLARENDON PRESS · OXFORD
1992

Oxford University Press, Walton Street, Oxford OX2 6DP
Oxford New York Toronto
Delhi Bombay Calcutta Madras Karachi
Petaling Jaya Singapore Hong Kong Tokyo
Nairobi Dar es Salaam Cape Town
Melbourne Auckland
and associated companies in
Berlin Ibadan

Oxford is a trade mark of Oxford University Press

Published in the United States
by Oxford University Press, New York

A catalogue record for this book is available from the British Library

Library of Congress Cataloging in Publication Data
Yeomans, J. M.
Statistical mechanics of phase transitions / J. M. Yeomans.
Based on a series of lectures given by the author at Oxford.
Includes bibliographical references.
ISBN 0–19–851729–7 (hbk.) ISBN 0–19–851730–0 (pbk.)
1. Phase transformations (Statistical physics)   I. Title.
QC175.16.P5Y46   1992   530.1'3—dc20   91-40516

Typeset by Pentacor PLC, High Wycombe, Bucks.
Printed in Great Britain by
Bookcraft (Bath) Ltd Midsomer Norton, Avon

Preface

The genesis of *Statistical mechanics of phase transitions* lies in a series of lectures I have given to physics graduates and undergraduates at Oxford over the past few years. I hope that it will be of use to future generations of students.

The book is also intended to act as, if not a bridge, a first stepping stone towards an understanding of phase transitions for those beginning research. By providing a summary of the field it may ease the first forays into the research literature.

Many scientists apart from theoretical physicists have an interest in phase transitions. I should be pleased if the book were read by experimentalists and researchers from other disciplines who would like to understand which theoretical approaches are available, when they can be expected to work, and why.

Particular thanks are due to Harvey Dobbs, Dr Philippe Binder, and Professor Eytan Domany for their helpful comments on the manuscript.

Oxford   J.M.Y
1991

Contents

1

Introduction

A phase transition occurs when there is a singularity in the free energy or one of its derivatives. What is often visible is a sharp change in the properties of a substance. The transitions from liquid to gas, from a normal conductor to a superconductor, or from paramagnet to ferromagnet are common examples.

The phase diagram of a typical fluid is shown in Fig. 1.1. As the temperature and pressure are varied water can exist as a solid, a liquid, or a gas. Well-defined phase boundaries separate the regions in which each state is stable. Crossing the phase boundaries there is a jump in the density and a latent heat, signatures of a first-order transition.

Consider moving along the line of liquid–gas coexistence. As the temperature increases the difference in density between the liquid and the gas decreases continuously to zero as shown in Fig. 1.2. It becomes zero at the critical point beyond which it is possible to move continuously from a liquid-like to a gas-like fluid. The difference in densities, which becomes non-zero below the critical temperature, is called the order parameter of the liquid–gas transition.

Seen on the phase diagram of water the critical point looks insignificant. However, there are clues that this might not be the case. Fig. 1.3 shows the specific heat of argon measured along the critical isochore, $\rho = \rho_c$. There is a striking signature of criticality: the specific heat diverges and is infinite at the critical temperature itself.

Analogous behaviour is seen in magnetic phase transitions. The phase diagram of a simple ferromagnet is shown in Fig. 1.4. Just as in the case of liquid–gas coexistence there is a line of first-order transitions ending in a critical point. All transitions occur at zero magnetic field, $H = 0$, because of the symmetry of a ferromagnet to reversals in the field. The additional symmetry means that it is often easier to work in magnetic language and we shall do so throughout most of this book.

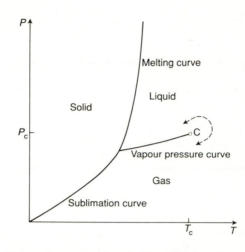

Fig. 1.1. Phase diagram of a fluid. All the phase transitions are first-order except at the critical point C. Beyond C it is possible to move continuously from a liquid to a gas. The boundary between the solid and liquid phases is thought to be always first-order and not to terminate in a critical point.

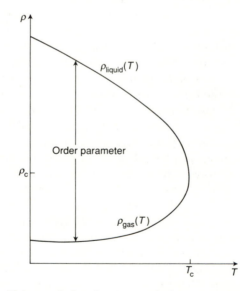

Fig. 1.2. Values of the densities of the coexisting liquid and gas along the vapour pressure curve. $(\rho_{liquid}(T) - \rho_{gas}(T))$ is the order parameter for the liquid–gas transition.

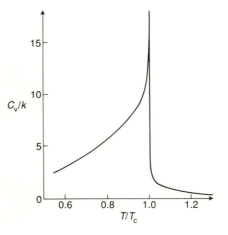

Fig. 1.3. Specific heat at constant volume of argon measured on the critical isochore, $\rho = \rho_c$. After Fisher, M.E. (1964). *Physical Review,* **136A**, 1599.

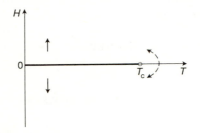

Fig. 1.4. Phase diagram of a simple ferromagnet. A line of first-order transitions at zero field ends in a critical point at a temperature T_c.

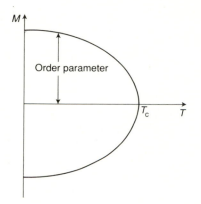

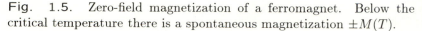

Fig. 1.5. Zero-field magnetization of a ferromagnet. Below the critical temperature there is a spontaneous magnetization $\pm M(T)$.

Crossing the phase boundary at temperatures less than the critical temperature, there is a jump in the magnetization. Above the critical temperature it is possible to move continuously from a state of negative magnetization to one of positive magnetization. The critical point itself separates these two behaviours; the magnetization is continuous but its derivatives are discontinuous. This manifests itself, just as in the fluid case, by divergences in the response functions, the specific heat and the susceptibility.

The order parameter for the ferromagnetic phase transition is the magnetization. Its variation with temperature along the coexistence curve, $H = 0$, is shown in Fig. 1.5. Compare this diagram with Fig. 1.2 for the fluid; the only difference is the extra symmetry in the magnetic case.

1.1   Phase transitions in other systems

Phase transitions in fluids and ferromagnets provide two simple examples of an enormous diversity of changes of state. Table 1.1 lists other examples, together with references for those wishing to pursue them further. We describe two cases in more detail to illustrate the richness and complexity of the phase diagrams found in nature.

1.1.1   A ferrimagnet: cerium antimonide

In cerium antimonide, strong uniaxial spin anisotropy constrains the spins to lie along the [100] direction. Within the (100) planes the

Table 1.1. Examples of the diversity of phase transitions found in nature

Transition	Example	Order parameter
ferromagnetic[a]	Fe	magnetization
antiferromagnetic[a]	MnO	sublattice magnetization
ferrimagnetic[a]	Fe_3O_4	sublattice magnetization
structural[b]	$SrTiO_3$	atomic displacements
ferroelectric[b]	$BaTiO_3$	electric polarization
order-disorder[c]	CuZn	sublattice atomic concentration
phase separation[d]	$CCl_4+C_7F_{16}$	concentration difference
superfluid[e]	liquid ^{4}He	condensate wavefunction
superconducting[f]	Al, Nb_3Sn	ground state wavefunction
liquid crystalline[g]	rod molecules	various

[a]Kittel, C. (1976). *Introduction to solid state physics* (6th edn). (Wiley, New York).

[b]Bruce, A. D. and Cowley, R. A. (1981). *Structural phase transitions.* (Taylor and Francis, London).

[c]Als-Nielsen, J. (1976). Neutron scattering and spatial correlation near the critical point. In *Phase transitions and critical phenomena*, Vol. 5a (eds C. Domb and M. S. Green), p.87. (Academic Press, London).

[d]Rowlinson, J. S. and Swinton, F. L. (1982). *Liquids and liquid mixtures* (3rd edn). (Butterworth Scientific, London).

[e]Wilks, J. and Betts, D. S. (1987). *An introduction to liquid helium* (2nd edn). (Clarendon Press, Oxford).

[f]McClintock, P. V. E., Meredith, D. J., and Wigmore, J. K. (1984). *Matter at low temperatures.* (Blackie, Glasgow and London).

[g]de Gennes, P.-G. (1974). *The physics of liquid crystals.* (Oxford University Press, Oxford).

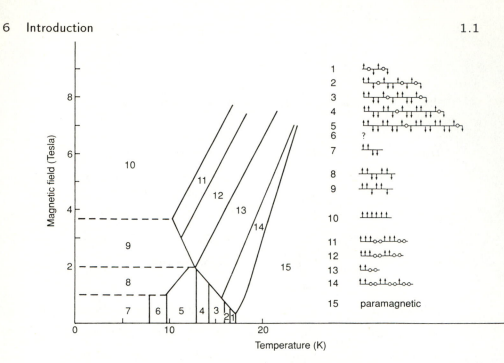

Fig. 1.6. The ferrimagnetic phases of cerium antimonide. The relative ordering of successive ferromagnetic planes in each phase is in-
dicated in the Figure. ○ denotes a plane with a net magnetization of
zero. After Rossat-Mignod, J., Burlet, P., Bartholin, H., Vogt, O.,
and Lagnier, R. (1980). *Journal of Physics C: Solid State Physics*, **13**,
6381, Institute of Physics Publishing Limited.

ordering is ferromagnetic: most planes lie in a state with spins $s = +1$
or $s = -1$, although planes with a net magnetization of zero are also
observed. The relative ordering of the planes themselves is ferrimagnetic. Fourteen different states, separated by first-order phase boundaries, have been identified in neutron scattering experiments. These
differ in the relative alignment of successive planes and are identified
in the phase diagram shown in Fig. 1.6. Note the patterns that link
the various sequences of phases: similar patterns are seen in series of
first-order transitions in binary alloys and minerals[1].

_____

[1]Yeomans, J.M. (1988). The theory and application of axial Ising
models. In *Solid state physics*, Vol. 41 (eds H. Ehrenreich, F. Seitz,
and D. Turnbull), p.151. (Academic Press, New York).

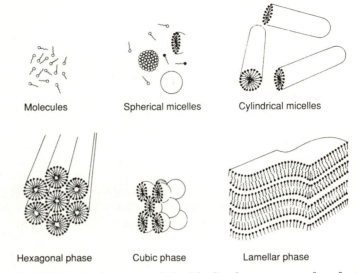

Fig. 1.7. Schematic drawings of the idealised structures of surfactant molecules that can form in solution as the surfactant concentration is increased. After Corkhill, J. M. and Goodman, J. F. (1969). *Advances in Colloid and Interface Science*, **2**, 297.

1.1.2   Surfactants in solution

Solutions of surfactant molecules have exotic phase diagrams[2]. These molecules have a polar head group which is very soluble in water and a hydrocarbon tail which is only just soluble. Hence they like to position themselves in such a way that the head is next to water molecules and the tail is shielded from them. If there is a surface they will migrate there and sit head-down. This lowers the surface tension—hence their use as soaps.

The phase diagrams of solutions of surfactant molecules are determined mainly by the concentration of the solute. As this increases micelles form. These are groups of molecules arranged in a sphere or cylinder so that the polar heads shield the hydrocarbon tails from the water. A further increase in concentration can lead to a phase transition to a state consisting of micelles ordered in a hexagonal or cubic array with the intervening spaces filled with water. A second transition is also observed in some systems. This is to a lamellar phase where the molecules are arranged into sheets but move freely within the sheets

---

[2]The future of industrial fluid design. In *Chemistry in Britain*, **26**, 4, April (1990).

like a two-dimensional liquid. Fig. 1.7 illustrates some of the possible phases.

Fluids, magnets, superconductors, surfactants: all apparently very different systems. Can the phase transitions associated with such diverse types of order be brought within the same theoretical framework? Why is there an order parameter, such as the magnetization, which becomes non-zero within the ordered phase? Why and how do the response functions diverge at the critical temperature? The aim of this book is to give an introduction to the theories that have been developed to answer these questions. A first step is to describe what is happening on a microscopic level at a phase transition with the aim of understanding the physics underlying the properties of a system at criticality.

1.2 A microscopic model

Consider a simple model of a two-dimensional interacting system, the Ising model on a square lattice. On each lattice site i there is a variable, called for convenience a spin, which can take two different values, $s_i = +1$ or $s_i = -1$. Each spin interacts with its nearest neighbours on the lattice through an exchange interaction, J, which favours parallel alignment

$$\mathcal{H} = -J \sum_{\langle ij \rangle} s_i s_j \qquad (1.1)$$

Fig. 1.8. A real-space renormalization group transformation for the two-dimensional Ising model on the square lattice. The initial configuration, corresponding to a temperature $T = 1.22T_c$, was generated using a Monte Carlo simulation. A sequence of renormalized configurations is then obtained by replacing successive clusters of nine spins by a single spin which takes the same value as the majority of the spins in the original cluster. Hence the length scale of the lattice is changed by a scale factor $b = 3$, 3^2, 3^3, and 3^4 in $(b),(c),(d)$, and (e) respectively. Note that the correlation length decreases under successive iterations of the renormalization group corresponding to an increase in the temperature. After Wilson, K. G. (1979). *Scientific American*, **241**, 140.

$T = 1.22\ T_c$

(a)

(b)

(c)   (d)   (e)

where we use the notation $\langle ij \rangle$ to represent a sum over nearest neighbour spins on sites i and j.

The two-dimensional Ising model has been solved exactly and is known to have a phase diagram like that shown in Fig. 1.4 with a continuous phase transition at zero field and a temperature T_c. The magnetization becomes non-zero at the critical temperature and increases to its saturation value, which corresponds to all the spins being aligned, at $T = 0$, just as in Fig. 1.5.

To see what is happening to individual spins as the temperature is changed it is not difficult to simulate the model on a computer with the fluctuations characteristic of finite temperatures being mimicked by a random number generator. This is the Monte Carlo method which will be described in more detail in Chapter 7. The results are shown in Figs 1.8–1.10. Black squares are used to represent spin $s_i = +1$ and white squares $s_i = -1$.

At temperatures very much greater than the critical temperature entropic contributions dominate the exchange energy and, although nearest neighbours tend to lie parallel, this is a small perturbation on a random configuration. Fig. 1.8(c) is an example of this. As the temperature is lowered the effects of the exchange interaction become more apparent. Nearest neighbours become more likely to point in the same direction and clusters of aligned or correlated spins appear. The size of the largest clusters is measured by a length called the correlation length. In Fig. 1.8(a) where the temperature is $1.2T_c$ the correlation length is of the order of a few lattice spacings. The system is said to show short-range order.

As the temperature is lowered the correlation length increases. Note, however, that fluctuations on a smaller scale remain important; there are correlated regions of spins on *all* length scales up to that set by the correlation length. Each fluctuation is not an area of uniform spin alignment but includes smaller fluctuations which in turn include yet smaller ones down to the length scale set by the lattice spacing . . .

> Clusters contain lesser ones
> Complicating quite 'em
> And lesser ones have lesser still
> Inside, *ad infinitum*.
> *(adapted from Jonathan Swift)*

The critical temperature itself is marked by the correlation length becoming infinite. A typical spin configuration at the critical temperature is shown in Fig. 1.9(a). There is now no upper length cut-off and ordered structures exist on every length scale. This is the microscopic

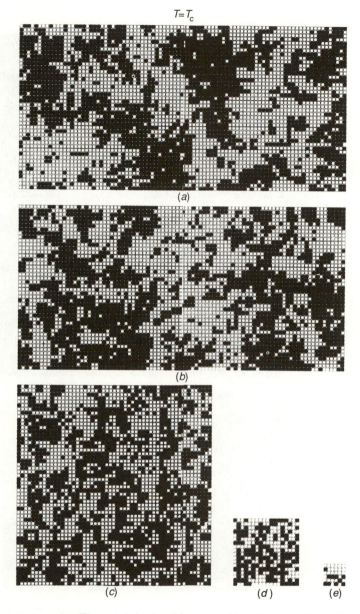

Fig. 1.9. As Fig. 1.8 but with a starting temperature $T = T_c$. Because the correlation length is initially infinite there is no change in the ordered state under iteration of the renormalization group and the system remains at the critical temperature. After Wilson, K. G. (1979). *Scientific American*, **241**, 140.

$T=0.99\ T_c$

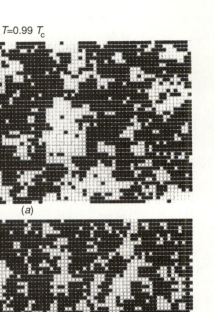

(a)

(b)

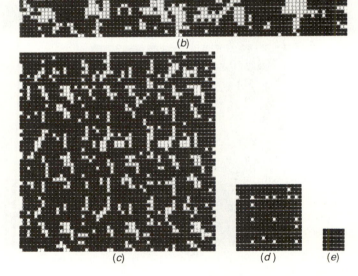

(c)   (d)   (e)

Fig. 1.10. As Fig. 1.8 but with a starting temperature $T = 0.99T_c$. Fluctuations relative to the ordered state are suppressed by the change in length scale and the system flows towards zero temperature. After Wilson, K. G. (1979). *Scientific American*, **241**, 140.

physics which underlies a critical phase transition. *Fluctuations on all scales of length are important.*

Below the critical temperature there is a non-zero magnetization. More spins lie in one of the two spin states: in Fig. 1.10 this is spin-up or black. The model is said to exhibit long-range order. At zero temperature all the spins are aligned because of the exchange interaction. As the temperature increases entropic terms in the free energy lead to fluctuations away from this state and the magnetization drops from its saturated value. Fig. $1.10(c)$ shows a spin configuration for a temperature $T \ll T_c$. The correlation length measures the size of the largest fluctuations away from the ordered background. As the temperature increases towards the critical temperature the correlation length becomes larger. Just as for $T > T_c$ there are clusters embedded within clusters on all length scales. The fluctuations cause the magnetization to fall, and it drops to zero exactly at the critical temperature where the correlation length becomes infinite and the underlying order is completely destroyed.

The long-range fluctuations in the magnetization of magnetic systems near the critical point are mirrored by long-range fluctuations in the density of fluid systems. These can be observed directly. If light is shone on to a fluid near its critical temperature it is reflected strongly, causing the fluid to appear milky-white. The strong scattering appears when the density fluctuations become of a size comparable to the wavelength of light, about a thousand times the interatomic spacing. This critical opalescence persists throughout the critical region emphazising that fluctuations at this length scale remain important even though the maximum length scale increases to infinity (mm or cm in a real sample).

1.2.1   A renormalization group

We have stressed that, at a critical point, all length scales are important. This is an unusual situation: usually physical theories can concentrate on a small range of scales of length. A continuum theory of water waves, ignoring atomic motions, or a theory of the arrangement of nucleons which ignores the atomic environment are essentially exact. So how can we cope with, or even exploit, scale invariance at criticality?

The answer lies in a set of theories known as renormalization groups. These will be described in much more detail in Chapters 8 and 9 but the ideas behind them can be illustrated using the Monte Carlo simulations in Figs 1.8–1.10. The aim is to change the scale of the system and see how it behaves. This is done by taking each group of nine spins in turn and replacing it by a single spin which takes the same value as the majority of spins in the original cluster. This procedure reduces the scale of the system by a factor $b = 3$. We then keep going to produce the series of snapshots of the spin configuration, essentially seen under different magnifications, shown in the figures.

For a starting temperature above the critical temperature (Fig. 1.8), the scale change soon obliterates any short-range order and the spins on the renormalized lattices become uncorrelated. This corresponds to an infinite temperature: the system has been renormalized by the simple transformation we have defined to $T = \infty$. This will be the case for all temperatures above T_c; the nearer to the critical temperature is the starting point the more steps of the transformation it will take to lose the short-range order.

For temperatures below the critical temperature there is an analogous flow as the renormalization group is iterated. However, now any fluctuations are relative to the ground state and, as these are lost under renormalization, the system flows to a completely ordered state characteristic of zero temperature. This is the case in Fig. 1.10.

Only at the critical temperature itself, Fig. 1.9, where there are fluctuations on all length scales does the system remain invariant under the renormalization group transformation. This can be exploited to identify the critical point and describe the behaviour of the thermodynamic functions in its vicinity.

2
Statistical mechanics and thermodynamics

This chapter moves through the large number of reminders and definitions necessary to arrive at the point where we can introduce the idea of universality, one of the most striking features of the theory of critical phenomena and a major justification for the interest in model systems. The first step is to summarize the statistical mechanics used throughout the book. Assuming that this is familiar material the main aim will be to gather together the relevant formulae in a form suitable for reference.

We then describe in more detail the behaviour of the thermodynamic functions at a phase transition, distinguishing between first-order and continuous transitions. It is very important to find a way of describing the asymptotic behaviour of these functions near a continuous transition and, to this end, we introduce the critical point exponents. A discussion of why they play a central role in the theory leads to the concept of universality.

2.1 Statistical mechanics

We assume that the reader is sufficiently familiar with elementary statistical mechanics to regard it as reasonable to start from the canonical partition function

$$\mathcal{Z}(T, H) = \sum_r e^{-\beta E_r} \tag{2.1}$$

where the sum is over all states r with energy E_r and $\beta = 1/kT$ with k Boltzmann's constant and T the temperature. Most of the subsequent chapters of this book will be concerned with models which, even if not applied to magnetic systems, are written in magnetic language, and therefore it is convenient to consider an ensemble in which $\mathcal{Z}$ depends on the temperature and the field H. Maxwell–Boltzmann statistics are appropriate because the magnetic systems we consider will consist of localized, and hence distinguishable, spins and the fluid systems will be in the classical regime.

The free energy is proportional to the logarithm of the partition function

$$\mathcal{F}(T, H) = -kT \ln \mathcal{Z}(T, H). \tag{2.2}$$

All macroscopic thermodynamic properties follow from differentiating the free energy. The relevant formulae are listed in Tables 2.1 and 2.2 for magnetic and fluid systems respectively. Readers unfamiliar with these should consult a text on statistical mechanics such as Callen[1]. Those who are rusty might find it helpful to try problems 2.1 and 2.2.

Often our aim will be to calculate the free energy. However, sometimes, particularly in numerical work, it is easier to extract properties such as the magnetization or the energy directly.

2.2 Thermodynamics

For a magnetic system the first law of thermodynamics can either be written[2]

$$dU = T\,dS - M\,dH \tag{2.3}$$

or

$$d\tilde{U} = TdS + HdM \tag{2.4}$$

where dU, dS, dH, and dM are the changes in the energy, entropy, magnetic field, and magnetization respectively. We have assumed the volume V is fixed and hence omitted the term $-PdV$. Both forms of the first law are equally valid but they correspond to different definitions of the energy. The energy stored in the applied magnetic field is not included in U, whereas it is included in $\tilde{U}$.

We shall use eqn (2.3) throughout because the free energy will then depend on the most convenient variables (T, H) and will be identical

---

[1]Callen, H. B. (1985). *Thermodynamics and an introduction to thermostatistics* (2nd edn). (Wiley, New York).

[2]The 'field', H, is taken to have the units of energy and the 'magnetization', M, to be dimensionless as is customary when writing spin Hamiltonians. If the field is the result of a *magnetic* field, B, they are related by $H \sim \mu_B B$ where μ_B is the Bohr magneton.

Table 2.1. The relation of the thermodynamic variables pertinent to a magnetic system to the partition function

---

Thermodynamic variables for a magnet

First law: $dU = T dS - M dH$

Partition function

$$\mathcal{Z}(T, H) = \sum_r e^{-\beta E_r}$$

$\downarrow$

Free energy

$$\mathcal{F} = -kT \ln \mathcal{Z}$$

Internal energy

$$U = -\frac{\partial \ln \mathcal{Z}}{\partial \beta}$$

Entropy

$$S = -\left(\frac{\partial \mathcal{F}}{\partial T}\right)_H$$
$$= (U - \mathcal{F})/T$$

Magnetization

$$M = -\left(\frac{\partial \mathcal{F}}{\partial H}\right)_T$$

Specific heat
(constant H)

$$C_H = \left(\frac{\partial U}{\partial T}\right)_H$$

Specific heat
(constant $X = H, M$)

$$C_X = T\left(\frac{\partial S}{\partial T}\right)_X$$

Isothermal susceptibility

$$\chi_T = \left(\frac{\partial M}{\partial H}\right)_T$$

Table 2.2. The relation of the thermodynamic variables pertinent to a fluid system to the partition function

---

Thermodynamic variables for a fluid

First law: $dU = TdS - PdV$

Partition function

$$\mathcal{Z}(T, V) = \sum_r e^{-\beta E_r}$$

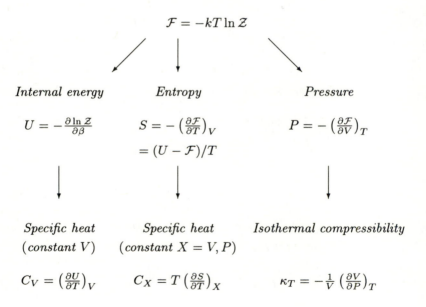

Free energy

$$\mathcal{F} = -kT \ln \mathcal{Z}$$

Internal energy	Entropy	Pressure
$U = -\frac{\partial \ln \mathcal{Z}}{\partial \beta}$	$S = -\left(\frac{\partial \mathcal{F}}{\partial T}\right)_V$ $= (U - \mathcal{F})/T$	$P = -\left(\frac{\partial \mathcal{F}}{\partial V}\right)_T$
Specific heat (constant V)	Specific heat (constant $X = V, P$)	Isothermal compressibility
$C_V = \left(\frac{\partial U}{\partial T}\right)_V$	$C_X = T\left(\frac{\partial S}{\partial T}\right)_X$	$\kappa_T = -\frac{1}{V}\left(\frac{\partial V}{\partial P}\right)_T$

to the function $\mathcal{F}$ defined in Section 2.1. To see this we recall that the thermodynamic definition of $\mathcal{F}$ is

$$\mathcal{F} = U - TS. \qquad (2.5)$$

Differentiating and using eqn (2.3)

$$d\mathcal{F} = dU - TdS - SdT = -MdH - SdT. \qquad (2.6)$$

Hence $\mathcal{F} \equiv \mathcal{F}(H, T)$. (It may avoid some confusion to note that if the alternative form of the first law (eqn 2.4) is used the free energy defined by eqn (2.5) becomes a function of M and T. This convention is used in some texts.)

2.3 Convexity properties of the free energy

A function $f(x)$ is a convex function of its argument x if

$$f(\frac{x_1 + x_2}{2}) \le \frac{f(x_1) + f(x_2)}{2} \qquad (2.7)$$

for all x_1 and x_2. If the inequality sign is reversed the function is said to be concave. A more useful definition for our purposes is that if the second derivative exists it must be ≥ 0 for a convex function and ≤ 0 for a concave function.

To determine the convexity properties of the free energy consider its second derivatives

$$\left(\frac{\partial^2 \mathcal{F}}{\partial T^2}\right)_H = \frac{-C_H}{T}; \qquad \left(\frac{\partial^2 \mathcal{F}}{\partial H^2}\right)_T = -\chi_T \qquad (2.8)$$

where C_H is the specific heat at constant field and χ_T is the isothermal susceptibility. It follows from the third law of thermodynamics that specific heats must be non-negative. Susceptibilities are usually positive, but there are exceptions, such as diamagnetic materials. However, it can be proved that if the Hamiltonian can be written

$$\mathcal{H} = \mathcal{H}_0 - HM \qquad (2.9)$$

they must be positive[3]. This formula will apply to all the cases which will be considered here. Because the second derivatives of the free

---

[3]Griffiths, R. B. (1965). *Journal of Chemical Physics*, **43**, 1958.

energy with respect to T and H are negative it is a concave function of both its variables.

2.4   Correlation functions

Thermodynamic variables like the magnetization or the entropy are macroscopic properties. In Section 1.2 it became apparent that a much fuller understanding of phase transitions could be obtained by considering what was happening on a microscopic level. To be able to do this in a more quantitative way we introduce correlation functions. For example the spin–spin correlation function, defined to measure the correlation between the spins on sites i and j, is

$$\Gamma(\vec{r}_i, \vec{r}_j) = \langle (s_i - \langle s_i \rangle)(s_j - \langle s_j \rangle) \rangle \tag{2.10}$$

where $\vec{r}_i$ is the position vector of site i and $\langle \ldots \rangle$ denotes a thermal average. If the system is translationally invariant $\langle s_i \rangle = \langle s_j \rangle$ and Γ depends only on $(\vec{r}_i - \vec{r}_j)$

$$\Gamma(\vec{r}_i - \vec{r}_j) \equiv \Gamma_{ij} = \langle s_i s_j \rangle - \langle s \rangle^2. \tag{2.11}$$

Away from the critical point the spins become uncorrelated as $r \to \infty$ and hence the correlation function decays to zero. Note that this is true not only above but also below the critical temperature, although here the mean value of the spin $\langle s \rangle \neq 0$, because, as is evident from eqn (2.10), the correlations are measured between the fluctuations of the spins away from their mean values. The correlations decay to zero exponentially with the distance between the spins

$$\Gamma(\vec{r}) \sim r^{-\tau} \exp^{-r/\xi} \tag{2.12}$$

where τ is some number. Equation (2.12) provides a definition of the correlation length, ξ, which was used in Section 1.2 as an estimate of the size of the largest ordered clusters in the Monte Carlo generated snapshots of an Ising model. We have assumed that ξ is independent of the direction of $\vec{r}$. This is usually the case for large r near criticality.

At the critical point itself long-range order develops in the system. The correlation length becomes infinite and eqn (2.12) breaks down. Evidence from experiments and exactly soluble models shows that here the correlation function decays as a power law

$$\Gamma(\vec{r}) \sim \frac{1}{r^{d-2+\eta}} \tag{2.13}$$

where η, our first example of a critical exponent, is a system-dependent constant[4].

It is possible to relate the spin–spin correlation function to the fluctuations in the magnetization and hence to the susceptibility. Using the formula relating the magnetization to the partition function given in Table 2.1 one can check that the fluctuations in the magnetization are given by

$$\langle (M - \langle M \rangle)^2 \rangle = \langle M^2 \rangle - \langle M \rangle^2 = k^2 T^2 \frac{\partial^2}{\partial H^2} \ln \mathcal{Z} = k T \chi_T. \quad (2.14)$$

But, writing the magnetization as a sum over spins,

$$\langle (M - \langle M \rangle)^2 \rangle = \sum_i (s_i - \langle s_i \rangle) \sum_j (s_j - \langle s_j \rangle) = \sum_{ij} \Gamma_{ij}. \quad (2.15)$$

For a translationally invariant system

$$\sum_{ij} \Gamma_{ij} = N \sum_i \Gamma_{i0} \sim N \int \Gamma(r) r^{d-1} dr \quad (2.16)$$

where the sum has been replaced by an integral, a step justified near criticality where the lattice structure is unimportant. Combining eqns (2.14), (2.15), and (2.16) we obtain

$$\chi_T \sim N \int \Gamma(r) r^{d-1} dr. \quad (2.17)$$

At the critical temperature the susceptibility diverges and hence $\Gamma(r)$ must become sufficiently long range that the integral on the right-hand side of eqn (2.17) also diverges. This sets an upper limit on η of 2. Note, from eqn (2.14), that a divergent susceptibility also implies a divergence in the fluctuations of the magnetization.

2.5 First-order and continuous phase transitions

A phase transition is signalled by a singularity in a thermodynamic potential such as the free energy. If there is a finite discontinuity in one or more of the first derivatives of the appropriate thermodynamic potential the transition is termed first-order. For a magnetic system the free energy $\mathcal{F}$, defined by eqn (2.5), is the appropriate potential

---

[4]Fisher, M. E. (1964). *Journal of Mathematical Physics*, **5**, 944.

with a discontinuity in the magnetization showing that the transition is first-order. For a fluid the Gibb's free energy, $\mathcal{G} = \mathcal{F} + PV$, is relevant and there are discontinuities in the volume and the entropy across the vapour pressure curve. A jump in the entropy implies that the transition is associated with a latent heat.

If the first derivatives are continuous but second derivatives are discontinuous or infinite the transition will be described as higher order, continuous, or critical[5]. This type of transition corresponds to a divergent susceptibility, an infinite correlation length, and a power law decay of correlations (eqn 2.13).

It will be helpful to look more carefully at how the thermodynamic variables behave near a phase transition for a particular case. The aim is to compare the behaviour at first- and higher order transitions and to look in some detail at the signatures of the latter with a view to defining the critical exponents in Section 2.6.

The example is the simple ferromagnet in a magnetic field. Its phase diagram was introduced in Chapter 1 and is reproduced for convenience in Fig. 2.1(a). There is a line of first-order transitions at zero field stretching from zero temperature to end at a critical point at a temperature $T = T_c$. The symmetry of the phase diagram, which is a consequence of the symmetry of a ferromagnet under reversals of the magnetic field, does not obscure any salient features. An example of a case where this symmetry is missing is the liquid–gas transition depicted in Fig. 1.1.

We first describe the field dependence of the free energy and its field derivatives, the magnetization and the susceptibility, along the three paths 1, 2, and 3 in Fig. 2.1(a). The aim is to compare the behaviour of these functions at temperatures below, equal to, and above T_c.

The free energy itself is shown in Fig. 2.1(b). Note that it is convex and symmetric about $H = 0$ as expected. A cusp develops at $H = 0$ for $T < T_c$. This signals a first-order phase transition as is seen more clearly in the behaviour of the magnetization, M.

The variation of M with H is shown in Fig. 2.1(c). For $T > T_c$ it varies continuously. For $T < T_c$, however, there is a jump at zero field indicative of the first-order phase transition. At the temperature di-

---

[5]The term 'second-order' phase transition, used synonymously with continuous phase transition, is a relic of the original classification of phase transitions into first-, second-, third- ... order due to Ehrenfest. This essentially recognized only discontinuities in thermodynamic derivatives, rather than divergences, which has been proved inappropriate. Therefore we follow M. E. Fisher in terming transitions first-order or continuous.

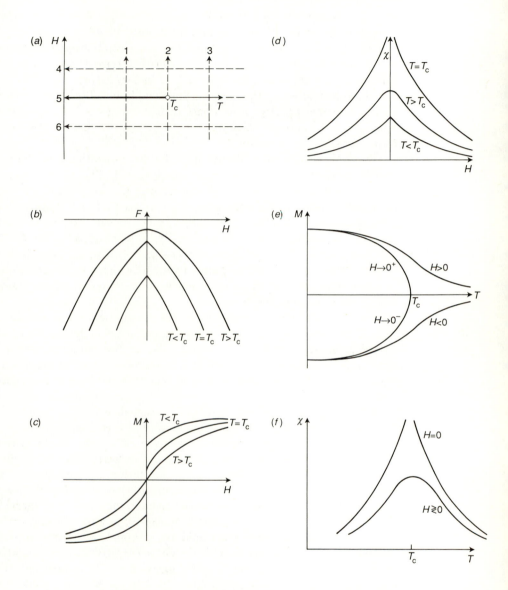

Fig. 2.1. (a) Phase diagram of a simple ferromagnet. There is a line of first-order transitions along $H = 0$ which ends at a critical point at $T = T_c$. (b) Field dependence of the free energy. (c) Field dependence of the magnetization. (d) Field dependence of the susceptibility. (e) Temperature dependence of the magnetization. (f) Temperature dependence of the susceptibility.

viding these behaviours, the critical temperature T_c, the magnetization is continuous at $H = 0$ but has infinite slope.

Differentiating again one obtains the isothermal susceptibility χ_T, which behaves in a definitive way at the critical temperature. The susceptibility is plotted as a function of field in Fig. 2.1(d). For $T > T_c$ it is a smooth function of the field as expected. Below T_c the susceptibility has a cusp at the first-order phase transition, $H = 0$. At the critical point itself the susceptibility diverges, a behaviour characteristic of a continuous phase transition.

We shall also be interested in how the magnetization and the susceptibility vary with temperature at constant field. This can be inferred from Figs 2.1(c) and 2.1(d) for the three paths 4, 5, and 6 in Fig. 2.1(a). Note that because of the symmetry of the magnetic phase diagram it is not possible to cross a line of first-order transitions by varying the temperature as would be the case generically. Following path 5 at $H = 0$ one passes through T_c and then follows a line of two-phase coexistence to zero temperature. Along paths 4 and 6, which have been chosen to lie equidistant from $H = 0$ to display the symmetry of the model better, there is no phase transition.

The temperature dependence of the magnetization is shown in Fig. 2.1(e). For non-zero field the magnetization increases smoothly with decreasing temperature to attain its saturation value, corresponding to all the spins being aligned, at zero temperature. The spins align along the direction of the field; if $H > 0$ the magnetization is positive and vice versa.

For $H = 0$ no preferred direction is singled out by the field and, for $T > T_c$, correlated regions of spins are finite and equally likely to point up or down. Hence the net magnetization is zero. At the critical temperature the correlation length becomes infinite, allowing a single cluster to dominate and a non-zero magnetization. The magnetization increases from zero at $T = T_c$ to its saturation value at $T = 0$. States with positive or negative magnetization have identical free energies. The two branches of the zero-field magnetization curve in Fig. 2.1(e) reflect this. The upper curve would be attained the presence of an infinitesimally small positive field; the curve corresponding to negative magnetization in an infinitely small negative field. Alternatively, cooling in a field and then taking the limit $H \rightarrow 0^+$ or $H \rightarrow 0^-$ would give positive or negative M respectively.

Finally we plot in Fig. 2.1(f) the susceptibility as a function of temperature. It must follow from symmetry that the susceptibility depends only on the magnitude of H, not on its sign. For finite field there is a peak in the susceptibility at T_c. For $H = 0$ this becomes a divergence signalling the critical point.

We have considered the dependence of the free energy on H and of its derivatives with respect to the field, the magnetization, and the susceptibility, on H and T. What about the temperature dependence of the free energy? For non-zero field there is no phase transition and hence the free energy is an analytic function of the temperature. For $H = 0$ one passes through a critical point as the temperature is lowered. This shows up in the second derivatives of the free energy.

Finally, for completeness, we mention the behaviour of the temperature derivatives of the free energy, the entropy, and the specific heat. At a first-order transition there is a usually a jump in the entropy and hence a latent heat[6]. The existence of a critical point is often marked by a specific heat which diverges at the critical temperature. An example of this is shown in Fig. 1.3.

2.6   Critical point exponents

We have argued that the critical point is marked by divergences in the specific heat and the susceptibility. It turns out to be very important to the theory of critical phenomena to understand more carefully the form of these divergences and the singular behaviour of the other thermodynamic functions near the critical point. To do this we define a set of critical exponents. We shall then start to justify why they play such a central role in the theory of critical phase transitions.

Let

$$t = (T - T_c)/T_c \tag{2.18}$$

be a measure of the deviation in temperature from the critical temperature T_c. Then the critical exponent associated with a function $F(t)$ is[7]

$$\lambda = \lim_{t \to 0} \frac{\ln |F(t)|}{\ln |t|} \tag{2.19}$$

or, as it is more usually written,

$$F(t) \sim |t|^\lambda . \tag{2.20}$$

The $\sim$ sign is well advised as it is important to remember that eqn (2.20) only represents the asymptotic behaviour of the function $F(t)$ as $t \to 0$. More generally one might expect

---

[6]For the ferromagnet the transition is between states of magnetization opposite in sign but equal in magnitude. Hence this is a transition with no associated latent heat.

[7]Assuming that the limit exists. See problem 2.3 for an example where this is not the case.

Table 2.3. Definitions of the most commonly used critical exponents for a magnetic system

Zero-field specific heat	$C_H \sim \mid t \mid^{-\alpha}$
Zero-field magnetization	$M \sim (-t)^\beta$
Zero-field isothermal susceptibility	$\chi_T \sim \mid t \mid^{-\gamma}$
Critical isotherm $(t = 0)$	$H \sim \mid M \mid^\delta \operatorname{sgn}(M)$
Correlation length	$\xi \sim \mid t \mid^{-\nu}$
Pair correlation function at T_c	$G(\vec{r}) \sim 1/r^{d-2+\eta}$

$$F(t) = A \mid t \mid^\lambda (1 + bt^{\lambda_1} + \ldots), \qquad \lambda_1 > 0. \qquad (2.21)$$

To check that this is a reasonable way of describing the leading behaviour of the singularities in the thermodynamic functions consider the zero-field magnetization of a ferromagnet shown in Fig. 2.1(e). Near T_c a sensible guess would be to describe the curve by a formula $M \sim (-t)^\beta$ with $\beta \sim 1/2$ because of the resemblance to a parabola.

The zero-field susceptibility diverges at T_c as shown in Fig. 2.1(f) and the zero-field specific heat shows qualitatively similar behaviour. Hence we may write

$$\chi_T \sim \mid t \mid^{-\gamma}; \qquad C_H \sim \mid t \mid^{-\alpha} \qquad (2.22)$$

where α and γ are positive.

A fourth exponent, δ, is introduced to describe the behaviour of the critical isotherm near the critical point at $H = 0$,

$$H \sim \mid M \mid^\delta \operatorname{sgn}(M) \qquad (T = T_c). \qquad (2.23)$$

Check that this corresponds to a curve of the form shown in Fig. 2.1(c). One might guess $\delta \sim 2$.

The critical exponent definitions are collected together in Table 2.3 for a magnetic system and Table 2.4 for a fluid. η and ν are associated with the pair correlation function and correlation length which were defined in Section 2.4. In particular, ν describes how the correlation length diverges as the critical temperature is approached.

Table 2.4. Definitions of the most commonly used critical exponents for a fluid system

Specific heat at constant volume V_c	$C_V \sim \mid t \mid^{-\alpha}$
Liquid–gas density difference	$(\rho_l - \rho_g) \sim (-t)^\beta$
Isothermal compressibility	$\kappa_T \sim \mid t \mid^{-\gamma}$
Critical isotherm $(t = 0)$	$P - P_c \sim$ $\mid \rho_l - \rho_g \mid^\delta \operatorname{sgn}(\rho_l - \rho_g)$
Correlation length	$\xi \sim \mid t \mid^{-\nu}$
Pair correlation function at T_c	$G(\vec{r}) \sim 1/r^{d-2+\eta}$

In compiling Tables 2.3 and 2.4 we have made the as yet totally unjustified assumption that the critical exponent associated with a given thermodynamic variable is the same as $T \to T_c$ from above or below. Early series and numerical work suggested that this was the case, but it was only with the advent of the renormalization group that it was indeed proved to be so. A common notation was to use a prime to distinguish the value of an exponent as $T \to T_c^-$ from the value as $T \to T_c^+$.

2.6.1   Universality

Having defined the critical exponents we need to justify why they are interesting. And indeed, why they are more interesting than the critical temperature T_c itself. It turns out that, whereas T_c depends sensitively on the details of the interatomic interactions, the critical exponents are to a large degree *universal* depending only on a few fundamental parameters. For models with short-range interactions these are the dimensionality of space, d, and the symmetry of the order parameter.

Striking evidence for this comes from a plot by Guggenheim presented as long ago as 1945. This is shown in Fig. 2.2 where the coexistence curves of eight different fluids are plotted in reduced units, T/T_c and ρ/ρ_c. Close to the critical point (and indeed surprisingly far away from it!) all the data lie on the same curve and hence can be described by the same exponent β. The fit assumes $\beta = 1/3$.

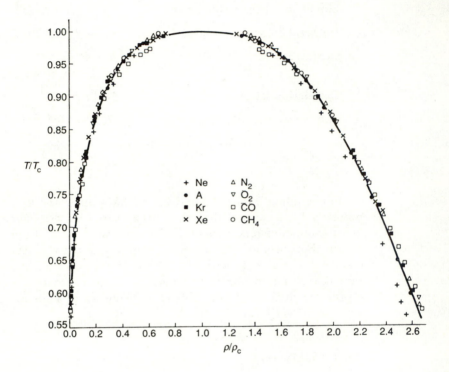

Fig. 2.2. The coexistence curve of eight different fluids plotted in reduced variables. The fit assumes an exponent $\beta = 1/3$. After Guggenheim, E. A. (1945). *Journal of Chemical Physics*, **13**, 253.

A further test of universality is to compare this value to that obtained for a phase transition in a completely different system with a scalar order parameter. Magnets with uniaxial anisotropy in spin space are one possibility—for MnF_2 a classic experiment by Heller and Benedek[8] gave $\beta = 0.335(5)$ where the number in brackets denotes the uncertainty in the final decimal place. For phase separation in the binary fluid mixture $CCl_4 + C_7F_{16}$ the experimental result[9] is $\beta = 0.33(2)$.

The Ising model, which we introduced as a simple example of an interacting system in Section 1.2 also has a scalar order parameter. It cannot be solved exactly in three dimensions but numerical estimates of the values of the critical exponents are very precise and provide a stringent test of universality. For the simple cubic, body-centred cubic, and face-centred cubic lattices $K_c = kT_c/J = 0.2216, 0.1574,$ and 0.1021 respectively. However, in all three cases β is the same, 0.327, with some argument about the value of the last decimal place[10].

This immediately illustrates the power of using simple models to describe critical behaviour. By making sure that one is working in the right dimension and that the symmetry of the order parameter is correctly represented by a model, it can be used to obtain critical exponents for *all* the systems within its universality class. It is much easier to study the Ising model than a complicated fluid Hamiltonian.

Universality classes are often labelled by the simplest model system belonging to them. Therefore a discussion of other universality classes will be postponed to the next chapter when we will have defined the relevant models.

2.6.2   Exponent inequalities

It is possible to obtain several rigorous inequalities between the critical exponents. The easiest to prove is due to Rushbrooke. It follows from the well known thermodynamic relation between the specific heats at constant field and constant magnetization

$$\chi_T(C_H - C_M) = T\left(\frac{\partial M}{\partial T}\right)_H^2. \qquad (2.24)$$

Because C_M must be greater than or equal to zero,

---

[8]Heller, P. and Benedek, G. B. (1962). *Physical Review Letters*, **8**, 428.

[9]Thompson, D. R. and Rice, O. K. (1964). *Journal of the American Chemical Society*, **86**, 3547.

[10]Liu, A. J. and Fisher, M. E. (1989). *Physica*, **A156**, 35.

$$C_H \geq T \left(\frac{\partial M}{\partial T}\right)_H^2 / \chi_T. \qquad (2.25)$$

As $t \to 0^-$ in zero field, using the definitions of the critical exponents in Table 2.3,

$$C_H \sim (-t)^{-\alpha}, \quad \chi_T \sim (-t)^{-\gamma}, \quad \left(\frac{\partial M}{\partial T}\right)_H \sim (-t)^{\beta-1}. \quad (2.26)$$

Therefore the inequality (2.25) can only be obeyed if

$$\alpha + 2\beta + \gamma \geq 2. \qquad (2.27)$$

Other inequalities, for example

$$\alpha + \beta(1 + \delta) \geq 2, \qquad (2.28)$$

can be obtained from the convexity properties of the free energy. Yet others, for example

$$\gamma \leq (2 - \eta)\nu; \quad d\nu \geq 2 - \alpha; \quad \gamma \geq \beta(\delta - 1), \qquad (2.29)$$

follow from making reasonable assumptions about the behaviour of the thermodynamic variables or correlation functions[11].

For the two-dimensional Ising model $\alpha = 0$, $\beta = 1/8$, $\gamma = 7/4$, $\delta = 15$, $\nu = 1$, and $\eta = 1/4$ and one can check that all the inequalities listed above actually hold as equalities. Exponents for some other universality classes are given in Table 3.1 and the reader might like to check whether the scaling laws are obeyed as equalities for these.

We have introduced two very new ideas, universality and inequalities between the critical exponents which appear to hold as equalities. The reader might well be demanding to know why the exponents have these striking properties. Such an explanation, based on the physics of scale invariance, will be forthcoming in Chapter 8 when the renormalization group is described. In the intervening chapters we look in more detail at models of systems which undergo phase transitions and how to calculate their critical exponents and other properties.

---

[11]The derivation of these inequalities is discussed in Stanley, H. E. (1971). *Introduction to phase transitions and critical phenomena*, Ch. 4. (Oxford University Press, Oxford).

2.7 Problems

2.1 (i) Verify eqn (2.14).

(ii) Show in a similar way that the fluctuations in the energy are related to the specific heat at constant volume by

$$(\Delta E)^2 \equiv \langle (E - \langle E \rangle)^2 \rangle = kT^2 C_V.$$

Use this equation to argue that $\Delta E \sim N^{1/2}$ where N is the number of particles in the system.

2.2 A paramagnetic solid contains a large number N of non-interacting, spin-1/2 particles, each of magnetic moment μ on fixed lattice sites. This substance is placed in a uniform magnetic field H.

(i) Write down an expression for the partition function of the solid, neglecting lattice vibrations, in terms of $x = \mu H/kT$.

(ii) Find the magnetization M, the susceptibility χ, and the entropy S, of the paramagnet in the field H.

(iii) Check that your expressions have sensible limiting forms for $x \gg 1$ and $x \ll 1$. Descibe the microscopic spin configuration in each of these limits.

(iv) Sketch M, χ, and S as a function of x.

[Answers: (i) $\mathcal{Z} = (2 \cosh x)^N$; (ii) $M = N\mu \tanh x$, $\chi = N\mu^2/(kT \cosh^2 x)$, $S = Nk\{\ln 2 + \ln(\cosh x) - x \tanh x\}$.]

2.3 Determine the critical exponents λ for the following functions as $t \to 0$:

$$
\begin{aligned}
&\text{(i)} && f(t) = At^{1/2} + Bt^{1/4} + Ct \\
&\text{(ii)} && f(t) = At^{-2/3}(t + B)^{2/3} \\
&\text{(iii)} && f(t) = At^2 e^{-t} \\
&\text{(iv)} && f(t) = At^2 e^{1/t} \\
&\text{(v)} && f(t) = A \ln\{\exp(1/t^4) - 1\}
\end{aligned}
$$

[Answers: (i)1/4, (ii)−2/3, (iii)2, (iv)undefined, (v)−4.]

2.4 Show that the following functions have a critical exponent $\lambda = 0$ in the limit $t \to 0$:

$$
\begin{aligned}
&\text{(i)} && f(t) = A \ln |t| + B \\
&\text{(ii)} && f(t) = A - Bt^{1/2} \\
&\text{(iii)} && f(t) = 1,\ t < 0; \qquad f(t) = 2,\ t > 0 \\
&\text{(iv)} && f(t) = A(t^2 + B^2)^{1/2}(\ln |t|)^2 \\
&\text{(v)} && f(t) = At \ln |t| + B
\end{aligned}
$$

2.5[12] Consider a model equation of state that can be written

$$H \sim aM(t + bM^2)^\theta; \quad 1 < \theta < 2; \quad a, b > 0.$$

near the critical point. Find the exponents β, γ, and δ and check that they obey the inequality given in (2.29) as an equality. [Answer: $\beta = 1/2$, $\gamma = \theta$, $\delta = 1 + 2\theta$.]

2.6[12] The spontaneous magnetization per spin of the spin-1/2 Ising model on the square lattice is

$$\langle s \rangle^8 = 1 - (\sinh 2J/kT)^{-4}.$$

Show that this can be written in the form

$$\langle s \rangle = B(-t)^\beta \{1 + b(-t) \ldots \}$$

where $t = (T - T_c)/T_c$ and $\beta = 1/8$. Find B and b and hence estimate the range of temperatures over which it is reasonable to ignore the correction to the leading scaling behaviour. [Answer: $B = (8\sqrt{2}K_c)^{1/8}$, $b = (1 - 9K_c/\sqrt{2})/8$ where $K_c = J/kT_c$.]

---

[12] After M. E. Fisher.

3

Models

The aim of this chapter is to describe some of the most fundamental models of cooperative behaviour. To model a physical system one route is to include, as realistically as possible, all the complicated many body interactions and try to obtain a quantitative prediction of the behaviour by solving Schrödinger's equation numerically. The other extreme is to write down the simplest possible model that still includes the essential physics and hope that it is tractable to analytic or precise numerical solution. The aim here is often to study universal behaviour or to gain a qualitative understanding of the physics governing the behaviour of a given class of materials.

It is the latter approach that we shall take here. Despite the apparent simplicity of the models, they show a rich mathematical structure and are in general difficult or, more usually, impossible to solve exactly. Moreover, and perhaps surprisingly at first sight, they do provide valid and useful representations of experimental systems. We shall return to discuss why this should be the case at the end of the chapter when armed with concrete examples.

It is conventional and convenient to use magnetic language and write the model Hamiltonians in terms of spin variables, although they will turn out to be applicable to many non-magnetic systems. In all the examples considered here the spins will lie on the sites i of a regular lattice. Three-dimensional lattices, such as simple cubic, body-centred cubic, and face-centred cubic, are familiar from conventional crystallography but we shall also be interested in lattices in two dimensions, such as the square, triangular, and hexagonal lattices shown in Fig. 3.1. and in one dimension where the lattice is just a linear chain of sites. It will become apparent in later chapters that most of the scientists in this field show a marked preference for working in any dimension but three.

33

(a) 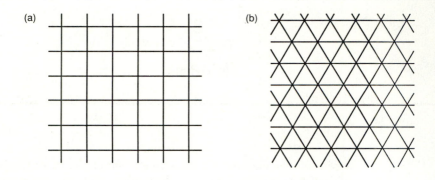 (b)

(c)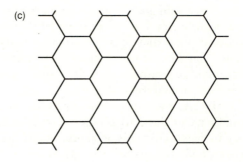

Fig. 3.1. Examples of two-dimensional regular lattices (a) square, (b) triangular, (c) hexagonal.

3.1 The **spin**-1/2 **Ising model**

A remarkably successful model of an interacting system, and one that
we shall use continually as an example throughout this book, is the
spin-1/2 Ising model. A classical spin variable s_i, which is allowed
to take values ± 1, is placed on each lattice site. The spins interact
according to a Hamiltonian

$$\mathcal{H} = -J \sum_{\langle ij \rangle} s_i s_j - H \sum_i s_i. \tag{3.1}$$

The first term in eqn (3.1) is responsible for the cooperative behaviour
and the possibility of a phase transition. J is the exchange energy:
positive J favours parallel and negative J antiparallel alignment of the
spins. We shall use $\langle ij \rangle$ to denote a sum over nearest neighbour spins;
further-neighbour interactions and terms which involve more than two
spins can be added to the Hamiltonian at will.

For $J = 0$, eqn (3.1) is the Hamiltonian of a paramagnet. A discussion of its statistical mechanics forms an early chapter in elementary
statistical mechanics texts. The only influence ordering the spins is
the field H. They do not interact, there are no cooperative effects and
hence no phase transition.

The Ising model is not difficult to solve in one dimension and we
shall do so (several times) as an example of the use of transfer matrices, series expansions and the renormalization group. However, one
dimension represents a special case because the phase transition is at
zero temperature.

The calculation of the exact partition function of the two-
dimensional Ising model in zero field was a mathematical *tour de force*
performed by Onsager in 1944. Extensions of his work mean that values are now known for all the critical exponents—they are rational
fractions in two dimensions for reasons that remained obscure for a
long time. The two-dimensional Ising model in a magnetic field and
the three-dimensional model, even in zero field, remain unsolved although their properties are known very precisely from numerical work.
Professor K. G. Wilson, who won the Nobel prize in 1982 for his work
on the renormalization group, describes:

When I entered graduate school I had carried out the instructions given to me
by my father and had knocked on both Murray Gell-Mann's and Feynman's
doors and asked them what they were currently doing. Murray wrote down
the partition function for the three-dimensional Ising model and said it would
be nice if I could solve it (at least that is how I remember the conversation).
Feynman's answer was 'nothing'.

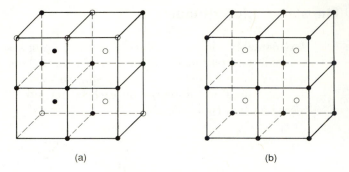

Fig. 3.2. A typical configuration of the copper and zinc atoms of beta-brass on the body-centred cubic lattice: (a) $T \gg T_c$; (b) $T \ll T_c$.

Despite its simplicity the Ising model is widely applicable because it describes any interacting two-state system. We illustrate this with two examples.

3.1.1 Order–disorder transitions in binary alloys

A classical example of a binary alloy is beta-brass. Beta-brass consists of equal numbers of copper and zinc atoms which lie on the sites of a body-centred cubic lattice. At high temperatures each lattice site is occupied at random by a copper or zinc atom giving the disordered structure shown in Fig. 3.2(a). We stress that the disorder is substitutional—the atoms occupy random positions on the lattice—rather than topological—the lattice itself has not ceased to exist, as would be the case for a liquid.

As the temperature is lowered there is, at $T_c = 733K$, a continuous phase transition to an ordered state where each atomic species preferentially occupies one of the two sublattices of the body-centred cubic lattice. The atomic configuration for $T \ll T_c$ is shown in Fig. 3.2(b). A suitable order parameter is the difference between the number of copper and zinc atoms on a chosen sublattice. Its variation with temperature is shown in Fig. 3.3.

Our aim is to write down a Hamiltonian which describes the interactions in beta-brass and predicts a continuous phase transition. To this end we assign the variables

$s_i = 1$ if site i is occupied by a copper atom,
$s_i = -1$ if site i is occupied by a zinc atom.

The spin on each site can take two values and hence is a spin-1/2 Ising variable. Defining J_{CuCu}, J_{ZnZn} and J_{CuZn} as the interaction

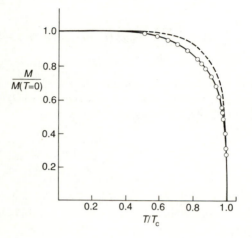

Fig. 3.3. Temperature dependence of the order parameter of beta-brass. The open circles are neutron scattering results, the dashed line X-ray scattering results, and the full line is the theoretical result for a compressible Ising model. The discrepancy between the X-ray and neutron data may arise because of the low sensitivity of X-rays to the atomic ordering. After Als-Nielsen, J. (1976). Neutron scattering and spatial correlation near the critical point. In *Phase transitions and critical phenomena*, Vol. 5a (eds C. Domb and M. S. Green), p.87. (Academic Press, London).

between two copper atoms, two zinc atoms, and a copper and a zinc atom respectively we may write the Hamiltonian

$$\mathcal{H} = \frac{1}{4}\sum_{\langle ij\rangle} J_{CuCu}(1+s_i)(1+s_j) + \frac{1}{4}\sum_{\langle ij\rangle} J_{ZnZn}(1-s_i)(1-s_j)$$

$$+ \frac{1}{4}\sum_{\langle ij\rangle} J_{CuZn}\{(1+s_i)(1-s_j) + (1-s_i)(1+s_j)\}. \qquad (3.2)$$

It is easy to check that if sites i and j are both occupied by copper atoms so that $s_i = s_j = 1$ this reduces to J_{CuCu} and so on. Collecting terms in eqn (3.2) gives

$$\mathcal{H} = -J\sum_{\langle ij\rangle} s_i s_j - H\sum_i s_i + C \qquad (3.3)$$

where $J = \frac{1}{4}(J_{CuCu} + J_{ZnZn} - 2J_{CuZn})$, C is a spin-independent term, and, because there are equal numbers of copper and zinc atoms, $\sum_i s_i = 0$.

We have arrived at the Hamiltonian of the nearest-neighbour spin-1/2 Ising model on a body-centred cubic lattice in zero field. What approximations are inherent in using this to describe beta-brass? Firstly I should like to stress that the use of an Ising variable is not an approximation (as long as there are no impurities or vacancies) as each lattice site is strictly in one of two states, occupied by copper or occupied by zinc. Therefore, because of the ideas of universality, the exponents should be those of the three-dimensional Ising model even if the details of the interatomic interactions are not well described by the Hamiltonian (3.2). This is borne out by the experimental values $\beta = 0.305 \pm 0.005$ and $\gamma = 1.24 \pm 0.015$[1] which should be compared to the current best estimates for the three-dimensional Ising model $\beta \approx 0.33$ and $\gamma \approx 1.24$.

To go beyond universal properties and try to predict experimental results like the variation of the order parameter with temperature the details of the interactions included in the model Hamiltonian become important. In general, further-neighbour interactions and multi-spin

---

[1] Als-Nielsen, J. (1976). Neutron scattering and spatial correlation near the critical point. In *Phase transitions and critical phenomena*, Vol. 5a (eds C. Domb and M. S. Green), p.87. (Academic Press, London). The discrepancy in β is thought to result from the thermal expansion of the lattice affecting the temperature dependence of the order parameter near criticality.

terms (such as $s_i s_j s_k$) and long-range interactions must be included to reproduce the thermodynamic functions correctly. In this particular example, however, they turn out to be unimportant. For beta-brass the most significant correction to the Ising model result comes from the variation of the exchange interaction J with temperature which results from the thermal expansion of the lattice. Allowing for this, the agreement with experiment is excellent, as shown in Fig. 3.3.

3.1.2 Lattice gas models

The archetypal lattice gas is a model where each lattice site can either be occupied by an atom or vacant. A variable $t_i = 1, 0$ is used to represent an occupied or unoccupied site respectively. The Hamiltonian is

$$\mathcal{H} = -J_L \sum_{\langle ij \rangle} t_i t_j - \mu_L \sum_i t_i \tag{3.4}$$

where J_L is a nearest neighbour interaction which favours neighbouring sites being occupied. μ_L is a chemical potential which controls the number of atoms: a large positive μ_L will lead to most sites being occupied whereas a large negative μ_L will favour vacancies.

As t_i is a two-state variable it must be possible to map it on to an Ising spin, $s_i = \pm 1$. This is achieved by the transformation

$$t_i = (1 - s_i)/2. \tag{3.5}$$

Substituting eqn (3.5) into eqn (3.4) one regains the usual nearest neighbour, spin-1/2 Ising Hamiltonian with the field related to the chemical potential.

A system which is well modelled by a lattice gas and which also illustrates the possibility of realizing experimental examples of the Ising model in *two* dimensions is hydrogen adsorbed on the (110) surface of iron. The atomic configuration of a (110) plane of iron is shown in Fig. 3.4(a). The potential wells between the iron atoms form a triangular lattice and define possible sites for the adsorption of hydrogen. Each site can either be occupied ($t_i = 1$) or vacant ($t_i = 0$) with the number of occupied sites, or coverage, being determined by the pressure of the hydrogen gas in contact with the surface.

As each adsorption site can be either occupied or vacant it has two states, and hence the phases of hydrogen on iron should be amenable to description by a lattice gas or equivalently an Ising model. As the coverage is varied several different ordered phases exist as the equilibrium state of the adsorbed hydrogen atoms. Some of these are shown in Fig. 3.4(b). They cannot be described by an Ising model with just

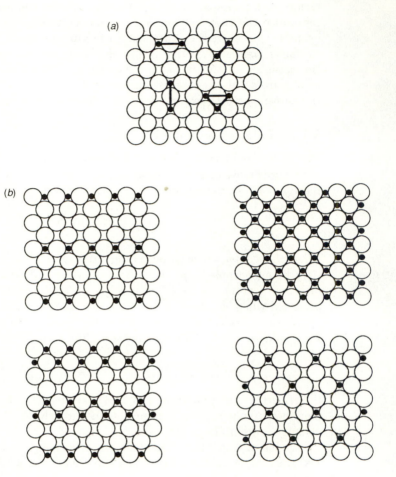

Fig. 3.4. (*a*) The atomic configuration of a (100) plane of iron showing the adsorption sites for the hydrogen atoms and the interactions
included in a model Hamiltonian for this system. (*b*) Some of the
resulting equilibrium phases.

nearest neighbour interactions, but by including the anisotropic second neighbour term and a three-spin interaction proportional to the product of spins around each elementary triangle shown in Fig. 3.4(a), the different phases and the transitions between them can be understood in some detail.

3.2 The spin-1 Ising model

For systems with more than two states higher-spin Ising models are appropriate. For example, the most general Hamiltonian for the spin-1 Ising model is

$$
\begin{aligned}
\mathcal{H} \quad = \quad & - J \sum_{\langle ij \rangle} s_i s_j - K \sum_{\langle ij \rangle} s_i^2 s_j^2 - D \sum_i s_i^2 \\
& - L \sum_{\langle ij \rangle} (s_i^2 s_j + s_i s_j^2) - H \sum_i s_i, \qquad s_i = \pm 1, 0. \quad (3.6)
\end{aligned}
$$

This follows from allowing all possible terms $s_i^\alpha s_j^\beta$; $\alpha, \beta = 0, 1, 2$. Higher powers of the spin do not enter because $s_i^3 = s_i$.

Because of its enlarged parameter space the spin-1 Ising model exhibits a much richer variety of critical behaviour than its spin-1/2 counterpart. The phase diagram for $K = L = 0$ is shown in Fig. 3.5. Three sheets of first order phase transitions join at a triple line where three phases coexist. The triple line ends in a tricritical point where the three phases become critical simultaneously.

3.3 The q-state Potts model

Many different spin models, some driven by theoretical and some by experimental considerations, have been defined in the scientific literature. Several examples appear in the problems at the end of this and subsequent chapters. The only other classical spin model that I shall define here is the q-state Potts model. The relation of this system to the physisorption of krypton atoms on a graphite surface provides an interesting example of how to construct a model Hamiltonian with the correct symmetry.

To define the Potts model a q-state variable, $\sigma_i = 1, 2, 3 \ldots q$, is placed on each lattice site. The interaction between the spins is described by the Hamiltonian

$$
\mathcal{H} = -J \sum_{\langle ij \rangle} \delta_{\sigma_i \sigma_j}. \qquad (3.7)
$$

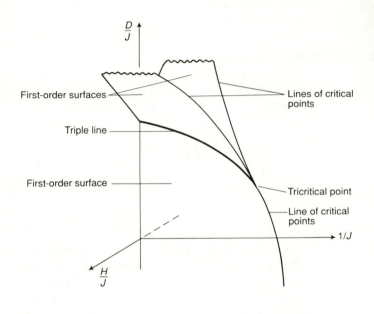

Fig. 3.5. A three-dimensional cross-section through the phase diagram of the spin-1 Ising model. Three surfaces of first-order transitions (two 'wings' and the lower portion of the $H = 0$ plane) meet at a triple line, shown in bolder type, where three phases coexist. The three phases become identical simultaneously at a tricritical point which marks the end of the triple line or, equivalently, the point where the three lines of critical points bounding the first-order surfaces meet.

δ is a Kronecker delta-function so the energy of two neighbouring spins is $-J$ if they lie in the same state and zero otherwise. It is easy to convince oneself that the Potts model has q equivalent ground states where all the spins are identical but can take any one of the q values. As the temperature is increased there is a transition to a paramagnetic phase which is continuous for $q \leq 4$ but first-order for $q > 4$ in two dimensions[2].

For $q = 2$ the Potts model is identical to the spin-1/2 Ising model. Note, however, that for $q = 3$ the Hamiltonian (3.7) does not correspond to the first term in eqn (3.6) because the three states of the spin-1 Ising model are not equivalent (see problem 3.2).

A physical realization of a system with the symmetry of the two-dimensional, three-state Potts model is krypton absorbed on the basal planes of graphite. The surface of graphite comprises hexagonal rings of carbon atoms and it is favourable for an adsorbed krypton to lie within one of the rings. However, the krypton atoms are sufficiently big that once a hexagon is occupied it becomes unfavourable for an atom to lie on any neighbouring site. Therefore, for one third coverage, the krypton atoms form a triangular lattice as shown in Fig. 3.6. But there are three entirely equivalent positions for the lattice: on the sublattices labelled a, b, and c in the figure. Hence the system has the symmetry of the three-state Potts model where a site corresponds to a triplet of adsorption rings and $\sigma_i = 1, 2, 3$ to the possibilities of the adsorbed krypton lying on the a, b, or c sublattices respectively.

3.4   X-Y and Heisenberg models

We have so far ignored the most obvious application of a spin model— to magnetic systems themselves. The restriction of the Ising model is that the spin vector can only lie parallel to the direction of quantization introduced by the magnetic field. This means that the Ising Hamiltonian can only prove useful in describing a magnet which is highly anisotropic in spin space. There are physical systems, MnF_2 for example, which to a good approximation obey this criterion, but fluctuations of the spin away from the axis of quantization must inevitably occur to some degree.

A more realistic model of many magnets with localized moments is

$$\mathcal{H} = -J_z \sum_{\langle ij \rangle} s_i^z s_j^z - J_\perp \sum_{\langle ij \rangle} (s_i^x s_j^x + s_i^y s_j^y) - H \sum_i s_i^z \qquad (3.8)$$

---

[2]Wu, F. Y. (1982). *Reviews of Modern Physics*, **54**, 235.

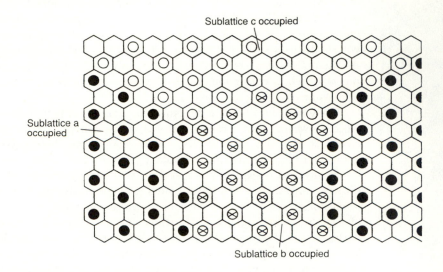

Fig. 3.6. Krypton adsorbed on the basal plane of graphite showing coexisting regions of the three ground states. After Kardar, M. and Berker, A. N. (1982). *Physical Review Letters*, **48**, 1552.

where x, y, and z label Cartesian axes in spin space. For $J_\perp = 0$ we regain the Ising model. For $J_z = J_\perp$ eqn (3.8) can be written

$$\mathcal{H} = -J \sum_{\langle ij \rangle} \vec{s}_i . \vec{s}_j - H \sum_i s_i^z. \tag{3.9}$$

This is the Heisenberg model.

The Heisenberg model was introduced in 1928 and was discussed in some detail as a model of ferromagnetism in Van Vleck's book of 1932[3]. It gives a reasonable description of the properties of some magnetic insulators, such as EuS, and provides a microscopic Hamiltonian describing the exchange interaction which leads to ferromagnetism. However, it does not include the possibility of non-localized spins and assumes complete isotropy in spin space.

The most fundamental theoretical difference between the Heisenberg and Ising models is that for the former the spin operators do not commute. Therefore it is a quantum mechanical rather than a classical

---

[3]Van Vleck, J. H. (1932). *The theory of electric and magnetic susceptibilities.* (Clarendon Press, Oxford).

spin model with corresponding greater difficulty in analytic or numerical treatments. Quantum models can be mapped on to classical spin systems in one higher dimension and there are some exact results for one-dimensional quantum models, just as for two-dimensional classical models[4]. Moreover, just as the Ising model only has a finite temperature phase transition for $d > 1$, the Heisenberg model orders at zero temperature unless $d > 2$.

The classical limit of the Heisenberg model can be constructed by taking the number of spin components to infinity and normalizing the spin from $\sqrt{S(S+1)}$ to 1. The spins become three-dimensional classical vectors. This limit, which leads to considerable simplifications in theoretical work, is useful because the critical exponents of the classical and quantum Heisenberg models are the same. This is an example of universality.

A second quantum mechanical spin model is the X-Y model, obtained by putting $J_z = 0$ in the Hamiltonian (3.8). This leads to spins which are two-dimensional, quantum mechanical vectors. The X-Y model, like the Heisenberg model, only has a conventional phase transition at non-zero temperature for $d > 2$. However, in $d = 2$ there is a transition at finite temperatures to an unusual ordered phase with quasi long-range order. This is marked by the correlations decaying algebraically (as in eqn 2.13) for all temperatures, not just at the critical point itself [5].

3.5   Universality revisited

In Section 2.6.1 the concept of the universality of critical exponents was described: that, for models with short-range interactions, the exponents depend only on the dimensionality of space and the symmetry of the order parameter. Several systems with the exponents of the three-dimensional Ising model were given as examples with the promise that more universality classes would be considered when the appropriate models had been introduced. We are now in a position to do this.

Universality classes which correspond to the models we have discussed in this chapter are listed in Table 3.1, together with an explicit description of the symmetry of the order parameter, physical examples, and the values of the critical exponents. This is a far from exhaustive

---

[4]Kogut, J. B. (1979). *Reviews of Modern Physics*, **51**, 659.

[5]Kosterlitz, J. M. and Thouless, D. J. (1978). Two-dimensional physics. In *Progress in Low Temperature Physics*, Vol VIIB (ed. D. F. Brewer), p.371. (North-Holland, Amsterdam).

Table 3.1. Universality classes

Universality class	Symmetry of order parameter	α	β	γ	δ	ν	η	Physical examples
2-d Ising	2-component scalar	0 (log)	1/8	7/4	15	1	1/4	some adsorbed mono e.g. H on Fe
3-d Ising	2-component scalar	0.10	0.33	1.24	4.8	0.63	0.04	phase separation, flu order-disorder e.g. β
3-d X-Y	2-dimensional vector	0.01	0.34	1.30	4.8	0.66	0.04	superfluids, supercon
3-d Heisenberg	3-dimensional vector	−0.12	0.36	1.39	4.8	0.71	0.04	isotropic magnets
mean-field		0 (dis.)	1/2	1	3	1/2	0	
2-d Potts, $q=3$ $q=4$	q-component scalar	1/3 2/3	1/9 1/12	13/9 7/6	14 15	5/6 2/3	4/15 1/4	some adsorbed mono e.g. Kr on graphite

list, but it includes many of the common experimental systems.

There are two questions which it is interesting to ask at this point, although a full explanation will not be forthcoming until later. Firstly, what universality class will a magnet that is neither strictly isotropic nor infinitely anisotropic, that is $J_\perp \neq J_z \neq 0$ in the Hamiltonian (3.8), belong to? This is the most common situation in reality.

It turns out that any anisotropy in the Hamiltonian, however weak, will eventually, as the system moves towards the critical temperature, drive the critical exponents away from Heisenberg values. The crossover temperature is determined by the strength of the anisotropy. If this is weak the critical behaviour will be Heisenberg-like over a wide range of temperatures and Ising or X-Y exponents may only be realized too close to the critical temperature to be experimentally observable. If the interaction J_z in the Hamiltonian (3.8) dominates, the exponents will cross over to Ising values; if $J_\perp$ is the stronger interaction the asymptotic critical behaviour will be X-Y like. Crossover is discussed further in Section 8.3.1.

A second point to note is that for dimensions $d \geq 4$ the exponents of the Ising, X-Y, and Heisenberg models become the same and take so-called mean-field values. The mean-field theories which are used to calculate these exponents are described in the next chapter. It is somewhat surprising that the exponents should suddenly lock into a dimensionality-independent value. The explanation of this will need the renormalization group. Note that as the dimensionality increases the Potts model (except for $q = 2$) does not show the same behaviour, but has a first-order transition. .

3.6   Discussion

In this chapter we have introduced several models and given examples of how they can describe experimental systems. We close by summarizing the importance of the approach of using simple models and discussing more generally why and to what extent they can give useful information about real systems whose behaviour is determined by complicated many-body interactions.

The prime advantage of using model systems is that they can be chosen to be tractable theoretically and therefore the details of their behaviour can be understood with some confidence. In particular, the aim is to extract a clear understanding of the physics leading to this behaviour which, it is hoped, will be mirrored in the real compounds.

Once the basic principles have been established various refinements or perturbations can be included in the models. Examples would be the effects of more complicated interactions, of defects, or of more realistic lattice structures. By ascertaining the robustness of the system to these perturbations it should be possible to establish whether they will significantly change the important physics and hence whether they are essential to model realistically a particular experimental system.

Often it is possible to go further than this and obtain a quantitative fit to experimental data, rather than just a qualitative understanding of it. For example, because critical exponents are universal and depend only on the dimensionality of space and the symmetry of the order parameter, a model has only to incorporate these properly to predict the correct critical behaviour.

It is often also feasible to obtain the behaviour of the thermodynamic functions throughout the whole range of temperature. This is because the interactions relevant to the physics under consideration can be mapped on to a few effective short-range terms. For example, for the case of hydrogen on iron, described in Section 3.1.2, where we are just looking at the ordering of the adsorbate, details of the iron–iron interactions are not important: they can just be considered to define a lattice of adsorption sites for the hydrogen atoms. Moreover, the complicated many-body interactions between the adsorbed atoms themselves can be well approximated by a simple spin Hamiltonian.

Exactly which interactions need to be included and their magnitudes must be confirmed by fitting to experimental results or by returning to a first principles calculation based on model atomic potentials. Hence a calculation of the critical temperature itself is in the realm of the band theorist and quantum chemist.

It is reassuring to be able to observe examples of spin models in nature. They also stand as interesting mathematical problems in their own right. How to study them forms the text of the remainder of this book.

3.7   Problems

3.1 Find the ground state (stable configuration at $T = 0$) of the following spin models:

(i) The one-dimensional Ising model with first and second neighbour interactions

$$\mathcal{H} = -J_1 \sum_i s_i s_{i+1} - J_2 \sum_i s_i s_{i+2}, \qquad s_i = \pm 1.$$

Consider both positive and negative values of the exchange parameters.

(ii) The one-dimensional, p-state chiral clock model

$$\mathcal{H} = -J \sum_i \cos\{2\pi(n_i - n_j + \Delta)/p\}, \quad n_i = 1, 2 \ldots p$$

for $J > 0$ and all values of Δ.

(iii) The spin-1 Ising model on a simple cubic lattice

$$\mathcal{H} = -J \sum_{\langle ij \rangle} s_i s_j - K \sum_{\langle ij \rangle} s_i^2 s_j^2 - D \sum_i s_i^2 \quad s_i = \pm 1, 0.$$

Consider both positive and negative values of the exchange interactions.

(iv) The antiferromagnetic spin-1/2 Ising model on a triangular lattice

$$\mathcal{H} = J \sum_{\langle ij \rangle} s_i s_j, \quad s_i = \pm 1$$

with $J > 0$.

3.2 Show that on the square lattice the spin-1 Ising model, descibed by the Hamiltonian (3.6), has the same symmetry as the three-state Potts model, described by the Hamiltonian (3.7), if

$$D + 2(J + K) = 0, \quad H = 0, \quad L = 0.$$

3.3 The one-dimensional, p-state clock model is described by the Hamiltonian

$$\mathcal{H} = -J \sum_{\langle ij \rangle} \cos\{2\pi(n_i - n_j)/p\}, \quad n_i = 1, 2 \ldots p.$$

Show that this model is equivalent to the q-state Potts model

$$\mathcal{H} = -J \sum_{\langle ij \rangle} \sigma_i \sigma_j, \quad \sigma_i = 1, 2 \ldots q$$

for $p = q = 2$ and $p = q = 3$ but not for higher values of p.

3.4 The Ising lattice gas is described by a Hamiltonian

$$\mathcal{H} = -J_L \sum_{\langle ij \rangle} s_i s_j t_i t_j - K_L \sum_{\langle ij \rangle} t_i t_j - D_L \sum_{(i)} t_i$$
$$s_i = \pm 1, \quad t_i = 0, 1.$$

Find a transformation which demonstates the equivalence of this model to the spin-1 Ising model defined by the Hamiltonian (3.6) with $H = L = 0$.

4

Mean-field theories

4.1 Mean-field theory for the Ising model

Very few statistical mechanical models have been solved exactly. In three dimensions not even the nearest neighbour spin-1/2 Ising model is tractable. Therefore it is necessary to resort to approximation methods. One of the most widely used of these is mean-field theory.

A systematic way of deriving the mean-field theory for a given microscopic Hamiltonian is to start from the Bogoliubov inequality[1]

$$\mathcal{F} \le \Phi = \mathcal{F}_0 + \langle \mathcal{H} - \mathcal{H}_0 \rangle_0 \tag{4.1}$$

where $\mathcal{F}$ is the true free energy of the system, $\mathcal{H}_0$ a trial Hamiltonian depending on a parameter H_0, $\mathcal{F}_0$ the corresponding free energy, and $\langle \dots \rangle_0$ denotes an average taken in the ensemble defined by $\mathcal{H}_0$.

The mean-field free energy is then defined by minimizing Φ with respect to the variational parameter H_0

$$\mathcal{F}_{mf} = min_{H_0}\{\Phi\}. \tag{4.2}$$

This gives the best possible approximation to the true free energy for a given choice of $\mathcal{H}_0$ because the inequality (4.1) insists that the mean-field free energy cannot fall below the true free energy. (This is analogous to the variational principle in quantum mechanics.) The usual choice for $\mathcal{H}_0$ is a free Hamiltonian—one with no interactions—as this enables explicit calculation of the right-hand side of (4.1).

As an example we consider the nearest neighbour Ising model in zero field, defined by the Hamiltonian (3.1) with $H = 0$, on a lattice

---

[1] A proof is given in Callen, H. B. (1985). *Thermodynamics and an introduction to thermostatistics* (2nd edn.), p.433. (Wiley, New York).

of N sites with each site having z nearest neighbours. For the simple cubic lattice $z = 6$. The trial Hamiltonian is

$$\mathcal{H}_0 = -H_0 \sum_i s_i, \tag{4.3}$$

where H_0 is termed the mean field.

This is just the Hamiltonian of a paramagnet. Therefore

$$\mathcal{F}_0 = -NkT \ln(2 \cosh \beta H_0), \tag{4.4}$$

$$\langle s \rangle_0 = \tanh \beta H_0. \tag{4.5}$$

We also need

$$
\begin{aligned}
\langle \mathcal{H} - \mathcal{H}_0 \rangle_0 &= \frac{\sum_{\{s\}} (-J \sum_{\langle ij \rangle} s_i s_j + H_0 \sum_i s_i) \exp[\beta H_0 \sum_i s_i]}{\sum_{\{s\}} \exp[\beta H_0 \sum_i s_i]} \\
&= -J \sum_{\langle ij \rangle} \langle s_i \rangle_0 \langle s_j \rangle_0 + H_0 \sum_i \langle s_i \rangle_0
\end{aligned} \tag{4.6}
$$

where the factorization of the interaction term is possible because $\mathcal{H}_0$ contains only single-site terms. For a translationally invariant system $\langle s_i \rangle_0 = \langle s_j \rangle_0 \equiv \langle s \rangle_0$ and the sums can be performed to give

$$\langle \mathcal{H} - \mathcal{H}_0 \rangle_0 = -JzN \langle s \rangle_0^2 / 2 + NH_0 \langle s \rangle_0 \tag{4.7}$$

where $zN/2$ is the number of bonds on the lattice ('each of N sites has z bonds each of which is shared between two neighbours!') Substituting eqns. (4.4) and (4.7) into (4.1) and using eqn (4.5) one obtains

$$\Phi = -NkT \ln(2 \cosh \beta H_0) - \frac{JzN}{2} \tanh^2 \beta H_0 + NH_0 \tanh \beta H_0. \tag{4.8}$$

Minimizing this expression with respect to H_0 gives a self-consistent expression for the mean field

$$H_0 = Jz \tanh \beta H_0 \tag{4.9}$$

or equivalently, noting from eqn (4.5) that $H_0 = Jz \langle s \rangle_0$, for the mean-field magnetization

$$\langle s \rangle_0 = \tanh \beta Jz \langle s \rangle_0. \tag{4.10}$$

Substituting eqn (4.9) back into eqn (4.8) shows that the mean-field free energy is

$$\mathcal{F}_{mf} = -NkT \ln(2 \cosh \beta Jz \langle s \rangle_0) + NJz \langle s \rangle_0^2 / 2. \tag{4.11}$$

To obtain the temperature dependence of the mean-field magnetization we need to solve eqn (4.10). The easiest way to obtain a qualitative

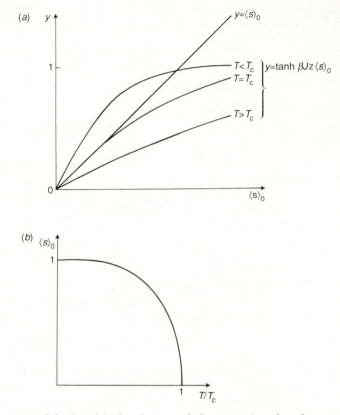

Fig. 4.1. (a) Graphical solution of the equation for the mean-field magnetization. (b) Temperature dependence of the mean-field magnetization.

understanding of how this behaves is to plot $\langle s \rangle_0$ and $\tanh(\beta J z \langle s \rangle_0)$ as functions of $\langle s \rangle_0$ for different values of the temperature. This is done in Fig. 4.1(a). The intersections of the curves correspond to values of $\langle s \rangle_0$ which solve eqn (4.10).

For temperatures greater than some value T_c, the only solution is $\langle s \rangle_0 = 0$ corresponding to the paramagnetic phase, whereas, for temperatures less than T_c, there are two solutions, $\langle s \rangle_0 = 0$ and a finite value of $\langle s \rangle_0$. One can check that the former corresponds to a maximum and the latter to a minimum of the mean-field free energy (4.11). Hence the stable phase is ferromagnetic. It is evident from Fig. 4.1(a) that the transition temperature T_c is calculated by equating the gradients of the two curves $\langle s \rangle_0$ and $\tanh(\beta J z \langle s \rangle_0)$ at the origin. This gives

$$kT_c = Jz. \tag{4.12}$$

Note that T_c depends only on z, the number of nearest neighbours, not on the other details of the lattice structure such as the dimensionality. One consequence of this is that mean-field theory incorrectly predicts a finite temperature phase transition for the one-dimensional Ising model.

Following the ferromagnetic solution as the temperature is decreased from T_c it is apparent from Fig. 4.1(a) that $\langle s \rangle_0$ increases continuously from zero and therefore that the phase transition is continuous. As the temperature tends to zero, $\langle s \rangle_0 \to 1$, because the tanh is bounded by unity. Physically this corresponds as expected to all the spins becoming aligned. $\langle s \rangle_0$ is plotted as a function of temperature in Fig. 4.1(b).

4.1.1   Mean-field critical exponents

As mean-field theory predicts a continuous transition for the Ising model it is possible to calculate the associated critical exponents, defined in Table 2.3. As a first example consider β, which describes the variation of the magnetization with temperature near T_c. It is helpful in calculations of this sort to make explicit the variables that are small. To this end, using the definition (2.18) and the value of the critical temperature given by eqn (4.12), we replace the temperature T by the reduced temperature t, which tends to zero at the critical point

$$T = T_c(1 + t) = \frac{Jz}{k}(1 + t). \tag{4.13}$$

Hence eqn (4.10) becomes

$$\langle s \rangle_0 = \tanh\{\langle s \rangle_0 / (1 + t)\}. \tag{4.14}$$

Expanding for small $\langle s \rangle_0$ and t:

$$\langle s \rangle_0 = \langle s \rangle_0 / (1 + t) - \langle s \rangle_0^3 / 3(1 + t)^3 + O(\langle s \rangle_0^5 / (1 + t)^5) \tag{4.15}$$

$$= \langle s \rangle_0 (1 - t) - \langle s \rangle_0^3 / 3 + O(\langle s \rangle_0 t^2, \langle s \rangle_0^3 t, \langle s \rangle_0^5). \tag{4.16}$$

Rearranging:

$$-t = \langle s \rangle_0^2 / 3 + O(t^2, \langle s \rangle_0^2 t, \langle s \rangle_0^4). \tag{4.17}$$

Therefore, if the correction terms can be ignored,

$$\langle s \rangle_0 \sim (-t)^{1/2}. \tag{4.18}$$

Note that t^2, $\langle s \rangle_0^2 t$, and $\langle s \rangle_0^4$ are all of order t^2 and it is indeed justified to neglect them. Hence the mean-field value of the exponent β, which we shall denote β_{mf}, is $1/2$.

The exponent α, which describes the temperature dependence of the specific heat, follows from differentiating the free energy twice. Hints as to the easiest way of doing this are given in problem 4.2. For $T < T_c$

$$C = \frac{3}{2}Nk + O(t). \tag{4.19}$$

For $T > T_c$

$$C = 0. \tag{4.20}$$

Thus there is a jump discontinuity in the specific heat and $\alpha_{mf} = 0$.

To obtain the exponents δ and γ we need to add a field term $-H\sum_i s_i$ to the Ising Hamiltonian. This will appear in the derivation in exactly the same way as H_0 so it follows immediately from eqns (4.9) and (4.10) that

$$\langle s \rangle_0 = \tanh \beta (Jz\langle s \rangle_0 + H). \tag{4.21}$$

For $T = T_c$, using eqn (4.12)

$$\langle s \rangle_0 = \tanh(\langle s \rangle_0 + H/Jz). \tag{4.22}$$

Expanding for small $\langle s \rangle_0$ and H

$$\langle s \rangle_0 = \langle s \rangle_0 + H/Jz - \langle s \rangle_0^3/3 + O(\langle s \rangle_0^2 H, \langle s \rangle_0 H^2, H^3, \langle s \rangle_0^5). \tag{4.23}$$

Therefore

$$\langle s \rangle_0 \sim H^{\frac{1}{3}} \tag{4.24}$$

and $\delta_{mf} = 3$. Check that the correction terms are indeed of lower order.

An outline of how to calculate the susceptibility exponent is given in problem 4.1. One obtains $\gamma_{mf} = 1$.

4.2   Landau theory

Landau theory is based on very simple assumptions, yet it not only predicts a phase transition but allows reproduction of the mean-field exponents showing very clearly how they depend on the symmetry of the order parameter. Landau assumed that the free energy can be expanded as a power series in the order parameter m where only those terms compatible with the symmetry of the system are included.

For example, for a simple ferromagnet in zero external field,

$$\mathcal{F} = \mathcal{F}_0 + a_2 m^2 + a_4 m^4 \tag{4.25}$$

because only even terms are invariant under a reversal in the sign of the magnetization. The series can be truncated after the term $O(m^4)$

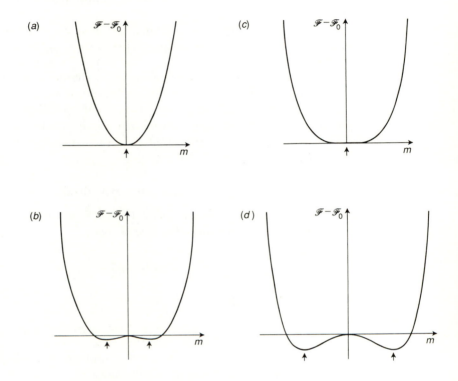

Fig. 4.2. Variation of the Landau free energy with magnetization for decreasing values of a_2. (a) $a_2 > 0$, (b) $a_2 = 0$, (c) $a_2 \lesssim 0$, (d) $a_2 < 0$.

because, as we shall see, if a_4 is taken to be positive, subsequent terms cannot alter the critical behaviour of the system.

The Landau free energy is plotted as a function of m for decreasing values of a_2 in Fig. 4.2. For $a_2 > 0$ the minimum value of $\mathcal{F}$ occurs at $m = 0$ showing that the system is in the paramagnetic phase. For $a_2 < 0$, however, the minimum free energy corresponds to a finite value of the order parameter m_0, indicating a ferromagnetic phase. Two stable states with magnetizations $+m_0$ and $-m_0$ coexist as a consequence of the physical symmetry built into $\mathcal{F}$. $a_2 = 0$ corresponds to the critical temperature where a spontaneous magnetization appears. Hence we may write

$$a_2 = \tilde{a}_2 t \qquad\qquad (4.26)$$

where t is the reduced temperature defined by eqn (2.18). Note that taking $a_4 > 0$ corresponds to the physical condition that the magnetization must be bounded.

It is at first sight surprising that singular behaviour can be obtained from a regular expansion. This happens because the value of the magnetization which minimizes the free energy is itself a singular function of the expansion coefficients which depend on the field and temperature. What is the order of the transition? Consideration of the shape of the free energy curve as a_2 decreases from zero (Figs 4.2(c) and 4.2(d)) shows that the magnetization becomes non-zero continuously but with a discontinuous derivative. Hence the transition is continuous and we should look for the critical exponents.

4.2.1   Mean-field critical exponents revisited

The easiest exponent to calculate is β. The equilibrium magnetization corresponds to the minimum of the free energy

$$\frac{d\mathcal{F}}{dm} = 2\tilde{a}_2 tm + 4a_4 m^3 = 0 \tag{4.27}$$

from which it follows immediately that, for $t < 0$,

$$m \sim (-t)^{1/2} \tag{4.28}$$

and $\beta_{mf} = 1/2$ where we anticipate in our notation that we shall regain the mean-field exponents derived in Section 4.1.1.

The specific heat exponent α follows from differentiating the free energy twice with respect to the temperature. Using eqn (4.27) to rewrite the expression (4.25) for $\mathcal{F}$ in terms of the reduced temperature t

$$\mathcal{F} = \mathcal{F}_0 - \frac{\tilde{a}_2^2 t^2}{4a_4} + O(t^3) \tag{4.29}$$

for $t < 0$. It is immediately apparent that the specific heat tends to a constant value as $t \to 0^-$. For $t > 0$ the equilibrium value of the magnetization is zero and therefore the specific heat is zero. Hence there is a jump discontinuity in the specific heat at the critical temperature and $\alpha_{mf} = 0$.

To obtain γ and δ we add a magnetic field term $-hm$ to the expression (4.25) for the free energy

$$\mathcal{F} = \mathcal{F}_0 - hm + \tilde{a}_2 tm^2 + a_4 m^4. \tag{4.30}$$

At equilibrium

$$\frac{d\mathcal{F}}{dm} = -h + 2\tilde{a}_2 tm + 4a_4 m^3 = 0. \tag{4.31}$$

On the critical isotherm $t = 0$ so

$$m^3 \sim h \tag{4.32}$$

and $\delta_{mf} = 3$. It also follows from 4.31 that $\gamma_{mf} = 1$ (see problem 4.5).

The exponents obtained from the Landau free energy (4.30) are the same as those calculated in Section 4.1.1 for the mean-field theory of the Ising model. It would be worrying if this were not the case because the Ising model has itself the symmetry with respect to reversals in the sign of the magnetization built into the Landau expansion. To see more explicitly why the same results are obtained we expand the mean-field free energy of the Ising model (4.11) for small $\langle s \rangle_0$

$$\mathcal{F}_{mf} = \mathcal{F}_0 + \frac{NJz}{2}\langle s \rangle_0^2 (1 - \beta Jz) + O(\langle s \rangle_0^4) \tag{4.33}$$

where the term in $\langle s \rangle_0^4$ is positive. This is the same as the expansion (4.25) and hence the critical exponents, which are determined by the behaviour for small magnetization, must be identical. Note that we may identify

$$a_2 = \frac{NJz}{2}(1 - \beta Jz) \tag{4.34}$$

and that the Landau expression for the critical temperature, $a_2 = 0$, is identical to eqn (4.12).

Indeed any model whose symmetry leads to a Landau expansion like (4.25) must have the same mean-field critical exponents. Ising models of higher spin, X-Y, and Heisenberg models are examples. But what of those cases where eqn (4.25) does not provide a satisfactory expansion? One example is when $a_4 < 0$. In this case a term $O(m^6)$ must be included. The resulting phase diagram, which includes both first-order and continuous transitions and a tricritical point, is explored in problem 4.6. If the symmetry of the order parameter allows a term in m^3 the transition is first-order. This is the case for the three-state Potts model and forms the thesis of problem 4.4.

4.3   The correlation function

To obtain the mean-field values of the exponents ν and η which describe the behaviour of the correlation function we need to invoke the

Ornstein–Zernike extension to Landau theory. This allows the magnetization to vary with position. We define a magnetization density $m(\vec{r})$ where $\vec{r}$ is a d-dimensional vector. To lowest order

$$\mathcal{F} - \mathcal{F}_0 = \tilde{a}_2 t \int \{m(\vec{r})\}^2 d^d \vec{r} + g \int \{\nabla m(\vec{r})\}^2 d^d \vec{r}. \qquad (4.35)$$

The first term on the right-hand side is just the quadratic term in eqn (4.25) written as an integral over d-dimensional space to allow for the $\vec{r}$-dependence of the magnetization. The second term is the lowest order term in the expansion of the spin–spin interaction and takes into account the extra free energy that results if the spins are not parallel.

The results we need can be obtained by working in Fourier space because, within the Ornstein–Zernike approximation, the modes corresponding to different wavevectors are independent of one another. The Fourier transform of $m(\vec{r})$, $\tilde{m}(\vec{q})$, is defined by

$$m(\vec{r}) = \frac{1}{(2\pi)^d} \int d^d \vec{q}\, e^{i\vec{q}\cdot\vec{r}} \tilde{m}(\vec{q}), \qquad (4.36)$$

$$\tilde{m}(\vec{q}) = \int d^d \vec{r}\, e^{-i\vec{q}\cdot\vec{r}} m(\vec{r}). \qquad (4.37)$$

Remembering that

$$\int d^d \vec{r}\, e^{-i\vec{q}\cdot\vec{r}} = (2\pi)^d \delta(\vec{q}) \qquad (4.38)$$

and that, because $m(\vec{r})$ is real, $\tilde{m}^*(\vec{q}) = \tilde{m}(-\vec{q})$ eqn (4.36) can be used to rewrite eqn (4.35) as

$$\mathcal{F} - \mathcal{F}_0 = \int \frac{d^d \vec{q}}{(2\pi)^d} (\tilde{a}_2 t + g q^2) \mid \tilde{m}(\vec{q}) \mid^2 . \qquad (4.39)$$

Equation (4.39) displays the free energy as a sum of quadratic terms, each of which, by equipartition of energy, contributes an average of kT

$$(\tilde{a}_2 t + g q^2)\langle \mid \tilde{m}(\vec{q}) \mid^2 \rangle = kT. \qquad (4.40)$$

The next step is to see how $\langle \mid \tilde{m}(\vec{q}) \mid^2 \rangle$ is related to the correlation function. This is defined by eqn (2.11) for discrete spins and we shall need the obvious generalization to continuous variables

$$\Gamma(\vec{r}) = \langle m(\vec{r})m(\vec{0}) \rangle - \langle m(\vec{0}) \rangle^2. \qquad (4.41)$$

Taking the Fourier transform of eqn (4.41), using eqns (4.36) and (4.37), and considering the paramagnetic phase[2] so that $\langle m(\vec{0})\rangle = 0$

$$\tilde{\Gamma}(\vec{q}) = \langle \tilde{m}(\vec{q})m(\vec{0})\rangle \tag{4.42}$$

$$= \int \frac{d^d\vec{q}'}{(2\pi)^d}\langle \tilde{m}(\vec{q})\tilde{m}(\vec{q}')\rangle \tag{4.43}$$

where we have neglected the $\vec{r}$-independent term. Because the different modes are uncorrelated

$$\langle \tilde{m}(\vec{q})\tilde{m}(\vec{q}')\rangle = \delta(\vec{q}+\vec{q}')(2\pi)^d\langle |\tilde{m}(\vec{q})|^2\rangle. \tag{4.44}$$

Using eqn (4.44) in eqn (4.43) the formula simplifies considerably to give

$$\tilde{\Gamma}(\vec{q}) = \langle |\tilde{m}(\vec{q})|^2\rangle. \tag{4.45}$$

Substituting eqn (4.40) into eqn (4.45) immediately gives an expression for the Fourier transform of the pair correlation function.

$$\tilde{\Gamma}(\vec{q}) = \frac{kT}{\tilde{a}_2 t + gq^2}. \tag{4.46}$$

Taking the inverse transform

$$\Gamma(\vec{r}) \sim \frac{e^{-r/\xi}}{r^{(d-3)/2}}, \qquad t \neq 0, \tag{4.47}$$

$$\Gamma(\vec{r}) \sim \frac{1}{r^{d-2}}, \qquad t = 0. \tag{4.48}$$

where $\xi = \sqrt{g/\tilde{a}_2 t}$. This means that $\nu_{mf} = 1/2$ and $\eta_{mf} = 0$.

4.4 Classical mean-field theories

Landau theory unifies a number of early theories of phase transitions which are now recognised as mean-field theories. These may well be familiar. We consider two of the most famous examples and point out their connection to the preceding discussion.

[2]The same exponents are obtained, as expected, for $T < T_c$. See Huang, K. (1987). *Statistical mechanics* (2nd edn.), p.425. (Wiley, New York).

4.4.1   Van der Waals theory of fluids

Van der Waals theory, which dates back to 1871, describes phase transitions in a fluid system. Van der Waals modified the ideal gas law by adding a term a/V^2 to the pressure to approximate the effects of intermolecular forces and replacing the volume V by $V - b$ to correct for the finite molecular size. Isotherms of the resulting equation

$$(P + \frac{a}{V^2})(V - b) = NkT \qquad (4.49)$$

are plotted in Fig. 4.3.

   The problem of what to do with the unstable $(dP/dV > 0)$ portions of the curve is circumvented by Maxwell's construction which allows two-phase coexistence: tie lines cutting equal areas above and below the curve join coexisting liquid and gas phases.

   The critical point, where the gas and liquid phases become identical, corresponds to a point of inflection of the isotherms

$$\left(\frac{\partial P}{\partial V}\right)_T = \left(\frac{\partial^2 P}{\partial V^2}\right)_T = 0. \qquad (4.50)$$

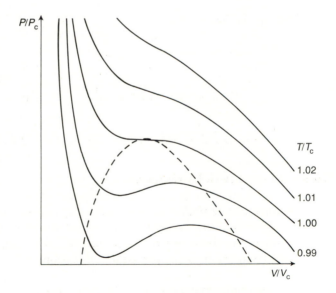

Fig. 4.3. Isotherms predicted by the Van der Waals theory of the liquid–gas phase transition.

Note that the second of these conditions is equivalent to a divergent compressibility. Using eqn (4.50) gives the critical parameters

$$P_c = a/27b^2, \quad V_c = 3b, \quad NkT_c = 8a/27b. \qquad (4.51)$$

Expanding the Van der Waals equation around the critical point gives mean-field values for the critical exponents. The equation is discussed further in problem 4.7.

4.4.2   Weiss theory of magnetism

The first theory of an interacting magnetic system was due to Weiss in 1907. He argued that the effects of cooperative ordering can be mimicked by assuming that each spin is subject to an effective field proportional to the magnetization. Hence in the usual equation for the magnetization of a paramagnet, eqn (4.5), the external field is supplemented by an internal field $\lambda\langle s \rangle_0$

$$\langle s \rangle_0 = \tanh \beta (H + \lambda \langle s \rangle_0) \qquad (4.52)$$

where λ is a constant which is related to the critical temperature, T_c. Note that eqn (4.52) is identical to eqn (4.21) if λ is identified as Jz. The derivation of the mean-field equations presented here, which starts from the Bogoliubov inequality (4.1), has allowed us to express the mean-field parameter λ in terms of the microscopic interaction J.

4.5   The validity of mean-field theory

We now want to address the questions of how far mean-field theory is correct and how far it is useful. A glance at Table 3.1 shows that the mean-field exponents do not agree with those of the two- or three-dimensional Ising or Heisenberg models or with those of real magnets or fluids. To see whether they ever provide the right answer we must look at what is wrong with mean-field theory.

The answer is that the theory ignores fluctuations: each spin is assumed to interact only with the mean of all the other spins in the system. This is immediately apparent from the Weiss formulation of mean-field theory described in Section 4.4.2. It is also inherent in the Landau theory where the free energy is expanded in terms of the magnetization, that is the mean value of the spins, rather than allowing an average over all spin configurations weighted by the appropriate Boltzmann factor.

So mean-field theory can only be valid when fluctuations are unimportant. We can estimate when this is the case by noting that the energy of a typical fluctuation is of order kT and its size, determined by the correlation length[3], is of order ξ^d. Therefore it gives a contribution to the free energy per unit volume of

$$\mathcal{F}_{fluc} \sim \frac{kT}{\xi^d} \sim |t|^{\nu d} \tag{4.53}$$

using $\xi \sim |t|^{-\nu}$. The specific heat, $C_H \sim |t|^{-\alpha}$. Integrating twice:

$$\mathcal{F} \sim |t|^{2-\alpha} . \tag{4.54}$$

For a consistent theory the free energy of a fluctuation must be much less than the total free energy. This implies that, as $t \to 0$,

$$d\nu_{mf} > 2 - \alpha_{mf} \tag{4.55}$$

or, putting $\alpha_{mf} = 0$ and $\nu_{mf} = 1/2$,

$$d_{mf} > 4. \tag{4.56}$$

Hence the mean-field theory is consistent and the exponents are correct if $d > 4$. $d = 4$ is called the upper critical dimension.

Different Landau expansions lead to different critical exponents and hence different upper critical dimensions. For example, for a tricritical point, the upper critical dimension is three. This is also the case for uniaxial Ising magnets with long range dipolar interactions, so mean-field exponents can be observed in the real world[3].

All models with the same Landau expansion share the same critical exponents. Hence mean-field theory leads to universality classes, although the degree of universality is overstated. In particular the exponents are independent of the dimensionality and the Ising, X-Y, and Heisenberg models share the same mean-field critical exponents. This ceases to be the case below the upper critical dimension.

Note that the heuristic argument given above concerns only the asymptotic behaviour of the thermodynamic functions as $T \to T_c$. Therefore, although the exponents are correct for $d > 4$, non-universal quantities such as transition temperatures are not. It is intuitively reasonable that mean-field theory, where the interaction is with an average of all the other spins in the system, should improve as the number of nearest neighbours or the range of the interactions increases. It is

---

[3]The reader who is worried about the effect of fluctuations on shorter length scales should consult Als-Nielsen, J. and Birgeneau, R. J. (1977). *American Journal of Physics*, **45**, 554.

not hard to prove that it becomes exact as the number of interacting neighbours z, the dimensionality, or the range of the interactions (if the limit is taken carefully) become infinite[4]. For example, for spin-1/2 Ising models, $T_c/T_c^{mf} = 1.76$ for the square lattice ($z = 4$), 1.33 for the simple cubic lattice ($z = 6$) and 1.23 for the face-centred cubic lattice ($z=12$).

The shortcomings of mean-field theory are often pointed out at great length but, although it is important to be aware of them, it puts things in better perspective to stress also the successes and wide applicability of the theory. Mean-field theory usually gives correct qualitative predictions for the phase diagrams of three-dimensional systems. It provides a feasible approach to complicated problems (often the only one), and it can be used as a starting point for more sophisticated calculations. Whereas many numerical approaches get harder as the dimensionality increases, mean-field theory improves. Finally, it highlights the importance of the symmetry of the order parameter.

4.6 Problems

4.1 Consider the spin-1/2 Ising model in a magnetic field

$$\mathcal{H} = -J \sum_{\langle ij \rangle} s_i s_j - H \sum_i s_i, \qquad s_i = \pm 1.$$

(i) Derive the self-consistent equation for the mean-field magnetization

$$\langle s \rangle_0 = \tanh \beta(Jz\langle s \rangle_0 + H)$$

where z is the coordination number of the lattice.
(ii) Differentiate this equation to show that the susceptibility per spin is given by

$$\chi = \frac{\partial \langle s \rangle_0}{\partial H} = \frac{(1 - \langle s \rangle_0^2)}{zJ(t + \langle s \rangle_0^2)}$$

where $t = (T - T_c)/T_c$ and T_c is the mean-field critical temperature.
(iii) Hence show that, for $t > 0$,

$$\chi = (zJt)^{-1}$$

---

[4]Thompson, C. J. (1988). *Classical equilibrium statistical mechanics*, p.95. (Clarendon Press, Oxford).

whereas, for $t < 0$,

$$\chi = (-2zJt)^{-1}$$

and therefore $\gamma_{mf} = 1$.

4.2 Obtain the mean-field value of the exponent α for the spin-1/2 Ising model. One approach is to
(i) Expand the mean-field free energy (eqn 4.11) for small $\langle s \rangle_0$ to terms $O(\langle s \rangle_0^4)$.
(ii) Write the expansion in terms of $t = (T - T_c)/T_c$ using $\langle s \rangle_0 = 0$ for $T > T_c$ and $\langle s \rangle_0 = (-3t)^{1/2}$ (see eqn 4.17) for $T < T_c$.
(iii) Differentiate twice with respect to the temperature to show that

$$C = 3Nk/2, \qquad T < T_c;$$
$$C = 0, \qquad T > T_c$$

and hence that $\alpha_{mf} = 0$.

4.3 Show that for a Hamiltonian

$$\mathcal{H} = -J \sum_{\langle ij \rangle} \vec{s}_i . \vec{s}_j$$

and a trial Hamiltonian

$$\mathcal{H}_0 = -\vec{H}_0 . \sum_i \vec{s}_i$$

the mean-field equations may be written

$$\vec{H}_0 = Jz\langle \vec{s} \rangle_0,$$

$$\mathcal{F}_{mf} = -kT \ln \mathcal{Z}_0 + JzN\langle \vec{s} \rangle_0^2/2$$

where the subscript 0 denotes an average in the ensemble defined by $\mathcal{H}_0$, N is the number of spins, and z is the coordination number of each spin.

4.4 The mean-field equations for the three-state Potts model

$$\mathcal{H} = -J \sum_{\langle ij \rangle} \delta_{\sigma_i \sigma_j}, \qquad \sigma_i = 1, 2, 3 \tag{4.57}$$

can be derived as follows using the result obtained in problem 4.3:
(i) Show that eqn 4.57 is equivalent to

$$\mathcal{H} = -J \sum_{\langle ij \rangle} \vec{s}_i . \vec{s}_j$$

where

$$\vec{s}_i = \begin{pmatrix} 1 \\ 0 \end{pmatrix}, \begin{pmatrix} -1/2 \\ \sqrt{3}/2 \end{pmatrix}, \begin{pmatrix} -1/2 \\ -\sqrt{3}/2 \end{pmatrix}.$$

(ii) Putting $\vec{H}_0 = \begin{pmatrix} H_0 \\ H_0' \end{pmatrix}$ show that the mean-field equations become

$$\frac{H_0}{Jz} = \frac{e^{3\beta H_0/2} - 1}{e^{3\beta H_0/2} + 2},$$

$$\mathcal{F}_{mf} = -NkT \ln(e^{\beta H_0} + 2e^{-\beta H_0/2}) + \frac{NH_0^2}{2Jz}.$$

(It is easiest to focus on an ordered state where $\begin{pmatrix} 1 \\ 0 \end{pmatrix}$ predominates so that $H_0' = 0$ by symmetry.)

(iii) Expand $\mathcal{F}_{mf}$ in $\langle s \rangle_0$ and show that it contains a cubic term. By sketching the mean-field free energy for suitable values of the coefficients in the expansion show that the transition is first-order.

(iv) Verify that the transition is at $kT_c = 3Jz/8 \ln 2$ and that the jump in the magnetization is $Jz/2$.

4.5 Consider a Landau expansion of the free energy

$$\mathcal{F} = \mathcal{F}_0 - hm + \tilde{a}_2 tm^2 + a_4 m^4.$$

By considering the variation of the equilibrium magnetization with the magnetic field show that the susceptibility is given by

$$\chi = (2\tilde{a}_2 t + 12a_4 m^2)^{-1}.$$

By writing m in terms of t for $t \to 0^+$ and $t \to 0^-$ show that the mean-field value of the susceptibility exponent γ_{mf} is unity and prove that

$$\frac{\chi(t \to 0^+)}{\chi(t \to 0^-)} = 2.$$

4.6 Consider a Landau expansion of the free energy of the form

$$\mathcal{F} = \frac{a}{2}m^2 + \frac{b}{4}m^4 + \frac{c}{6}m^6, \quad c > 0.$$

Prove that there is a line of critical transitions $a = 0$, $b > 0$ which joins a line of first order transitions $b = -4(ca/3)^{1/2}$ at a tricritical point $a = b = 0$. Sketch the form of the free energy in each region of the (a, b) plane, on the transition lines, and at the tricritical point[5].

4.7 (i) Show that the critical parameters of the Van der Waals equation of state for a fluid (eqn 4.49) are

$$P_c = a/27b^2, \quad V_c = 3b, \quad NkT_c = 8a/27b.$$

(ii) Hence show that, when written in terms of reduced variables

$$p = P/P_c, \quad v = V/V_c, \quad t = T/T_c,$$

the equation takes the universal form

$$(p + 3/v^2)(v - 1/3) = 8t/3.$$

This is the law of corresponding states. Although the quantitative form of the equation is incorrect for fluids in three dimensions near the critical point, the idea of an equation of state which, when written in reduced variables, is universal has been very important in the development of the theory of critical phenomema. See Fig. 2.2 for experimental evidence.
(iii) Obtain values for the critical exponents β, γ, δ of the Van der Waals theory and confirm that they take mean-field values[6].

---

[5]For an analysis which includes the terms in odd powers of m see Appendix A of Sarbach, S. and Fisher, M. E. (1979). *Physical Review*, **B20**, 2797.
[6]To obtain a value for α requires knowledge of the free energy. See Thompson, C. J. (1988). *Classical equilibrium statistical mechanics*, p.84. (Clarendon Press, Oxford).

5

The transfer matrix

The aim of this chapter is to describe how transfer matrices can be used to solve one-dimensional classical spin models. The idea is to write down the partition function in terms of a matrix, the transfer matrix. The thermodynamic properties of the model are then wholly described by the eigenspectrum of the matrix. In particular the free energy per spin in the thermodynamic limit depends only on the largest eigenvalue and the correlation length only on the two largest eigenvalues through simple formulae.

The simplest application of the transfer matrix technique is to the exact solution of one-dimensional spin models with a finite number of neighbours per site and a finite number of spin states. Transfer matrices have, however, also proved very useful in the solution of exactly solvable two-dimensional models; now the matrices are infinite-dimensional and their analysis requires sophisticated mathematics[1].

5.1 Setting up the transfer matrix

We shall use the one-dimensional Ising model in a magnetic field as an explicit example of how to set up a transfer matrix. This model is described by the Hamiltonian

$$\mathcal{H}_N = -J \sum_{i=0}^{N-1} s_i s_{i+1} - H \sum_{i=0}^{N-1} s_i \tag{5.1}$$

where we shall, for convenience, take periodic boundary conditions, that is identify $s_N \equiv s_0$. The choice of boundary conditions becomes irrelevant in the thermodynamic limit, $N \to \infty$.

---

[1]Baxter, R. J. (1982). *Exactly solved models in statistical mechanics.* (Academic Press, London and San Diego).

The partition function, written out in some detail, is

$$Z = \sum_{\{s\}} e^{\beta J(s_0 s_1 + s_1 s_2 + \dots + s_{N-1} s_0) + \beta H(s_0 + s_1 + \dots + s_{N-1})}$$

(5.2)

where $\{s\}$ represents the trace over all possible states of the system, that is the sum over $s_i = \pm 1$ for all spins s_i. The important property of eqn (5.2) that allows it to be represented as a product of matrices is that it can be rearranged into products of terms each depending only on nearest neighbour pairs

$$Z = \sum_{\{s\}} e^{\beta J s_0 s_1 + \beta H(s_0 + s_1)/2} \, e^{\beta J s_1 s_2 + \beta H(s_1 + s_2)/2} \dots$$

$$\dots e^{\beta J s_{N-1} s_0 + \beta H(s_{N-1} + s_0)/2}$$

(5.3)

$$\equiv \sum_{\{s\}} \mathbf{T}_{0,1} \mathbf{T}_{1,2} \dots \mathbf{T}_{N-1,0}$$

(5.4)

where

$$\mathbf{T}_{i,i+1} = e^{\beta s_i s_{i+1} + \beta H(s_i + s_{i+1})/2}$$

(5.5)

are the elements of a matrix $\mathbf{T}$ with rows labelled by the values of s_i and columns by the values of s_{i+1}. Writing out $\mathbf{T}$ explicitly for the model we are considering

$$\begin{matrix} & s_{i+1} = 1 & s_{i+1} = -1 \\ \begin{matrix} s_i = 1 \\ s_i = -1 \end{matrix} & \begin{pmatrix} e^{\beta(J + H)} & e^{-\beta J} \\ e^{-\beta J} & e^{\beta(J - H)} \end{pmatrix} \end{matrix}$$

(5.6)

Equation (5.4) is easily simplified by noting that it is a matrix product written in terms of the components of the matrix $\mathbf{T}$. Taking the trace over the spins $i = 1, 2, \dots, N - 1$ corresponds to performing the product

$$Z_N = \sum_{s_0 = \pm 1} (\mathbf{T}^N)_{0,0}$$

(5.7)

so that only the summation over s_0 of the diagonal elements of $\mathbf{T}^N$ remains. This is just the trace of $\mathbf{T}^N$ which is most usefully expressed in terms of the eigenvalues λ_i of $\mathbf{T}$

$$Z_N = \sum_i \lambda_i^N.$$

(5.8)

Although we have used the example of the one-dimensional Ising model to enable us to display an explicit formula at each step, eqn (5.8) is a general result.

The transfer matrix method is useful whenever the partition function can be factorized in a form like eqn (5.3) and hence expressed as a product of matrices. A common application is to one-dimensional classical spin systems with finite-range interactions. The size of the transfer matrix depends on the number of spin states per site and on the range of the interactions. For example, for the nearest neighbour q-state Potts model it is $q \times q$. For the one-dimensional Ising model with first and second neighbour interactions the rows and columns are labelled by s_i, s_{i+1} and s_{i+2}, s_{i+3} respectively and hence the matrix is 4×4. As the model gets more complicated the usefulness of the formalism depends on whether the transfer matrix can be diagonalized analytically or numerically.

A pictorial way of thinking of the transfer matrix is that it builds up the lattice step by step. Multiplying by the Rth power of $\mathbf{T}$ adds the spin s_R and traces over the spin s_{R-1}. Hence this step can be considered to add the bond between spins $R-1$ and R. Any further terms in $\mathcal{Z}$ cannot depend on the value of s_{R-1} as the trace has already been taken over this spin.

5.2 The free energy

The power of the transfer matrix formalism becomes apparent in the formula for the free energy. We shall now leave the example of the Ising model and consider a general transfer matrix $\mathbf{T}$ of size $n \times n$. If the eigenvalues, listed in terms of decreasing modulus, are labelled $\lambda_0, \lambda_1, \lambda_2 \ldots \lambda_{n-1}$ then, in the thermodynamic limit, the free energy per spin is given by

$$f \;=\; -kT \lim_{N \to \infty} \frac{1}{N} \ln \mathcal{Z}_N \tag{5.9}$$

$$=\; -kT \lim_{N \to \infty} \frac{1}{N} \ln \left\{ \lambda_0^N \left(1 + \sum_i \frac{\lambda_i^N}{\lambda_0^N} \right) \right\}. \tag{5.10}$$

But, as $N \to \infty$, $(\lambda_i/\lambda_0)^N \to 0$ because the ratio is less than 1 and hence

$$f = -kT \ln \lambda_0. \tag{5.11}$$

This is an important result because it is often much easier to calculate λ_0 than the entire spectrum of a matrix.

It is not necessary to worry about degeneracy in λ_0 because transfer matrices can be proved to belong to a class of matrices with non-degenerate, positive largest eigenvalue λ_0, thus giving a physically sensible free energy[2]. We have assumed that the λ_i are real. This is not necessarily the case for $i \neq 0$ but the formula (5.11) still holds (see problem 5.3).

5.3   The correlation function

A second important quantity which is simply related to the eigenvalues of the transfer matrix is the correlation length. To calculate this we need the spin–spin correlation function which serves as an example of how to obtain averages of products of spins using transfer matrices. We recall from Chapter 2 the definitions of Γ_R, the two-spin correlation function, and ξ, the correlation length,

$$\Gamma_R = (\langle s_0 s_R \rangle - \langle s_0 \rangle \langle s_R \rangle), \tag{5.12}$$

$$\xi^{-1} = \lim_{R \to \infty} \left\{ -\frac{1}{R} \ln | \langle s_0 s_R \rangle - \langle s_0 \rangle \langle s_R \rangle | \right\}. \tag{5.13}$$

Consider first the calculation of

$$\langle s_0 s_R \rangle_N = \frac{\sum_{\{s\}} s_0 s_R e^{-\beta \mathcal{H}_N}}{\sum_{\{s\}} e^{-\beta \mathcal{H}_N}} \equiv \frac{1}{\mathcal{Z}_N} \sum_{\{s\}} s_0 s_R e^{-\beta \mathcal{H}_N} \tag{5.14}$$

where the subscript N denotes that we are again considering a ring of N spins. $\mathcal{Z}_N$ is known from eqn (5.8) and the numerator can be written in a form analogous to eqn (5.4)

$$\sum_{\{s\}} s_0 s_R e^{-\beta \mathcal{H}_N} = \sum_{\{s\}} s_0 \, \mathbf{T}_{0,1} \mathbf{T}_{1,2} \ldots \mathbf{T}_{R-1,R} \, s_R \, \mathbf{T}_{R,R+1} \ldots \mathbf{T}_{N-1,0}$$

$$= \sum_{s_0 s_R} s_0 \, (\mathbf{T}^R)_{0,R} \, s_R \, (\mathbf{T}^{N-R})_{R,0}. \tag{5.15}$$

Let $\mathbf{T}$ have eigenvectors $| \, \vec{u}_i \rangle$ corresponding to the eigenvalues λ_i, $i = 0, 1, 2, \ldots n - 1$. It will also be useful to define the diagonal matrix $\mathbf{s}_R$

---

[2]This is the Perron–Frobenius theorem which is discussed in Horn, R. A. and Johnson, C. A. (1985). *Matrix analysis*, p.508. (Cambridge University Press, Cambridge).

with eigenvalues equal to the possible values of s_R and corresponding eigenvectors $\langle \vec{s}_R \mid = (00 \ldots 010 \ldots 00)$. Making use of the formulae

$$\mathbf{s}_R = \sum_{s_R} \mid \vec{s}_R \rangle s_R \langle \vec{s}_R \mid, \qquad (5.16)$$

$$\mathbf{T} = \sum_i \mid \vec{u}_i \rangle \lambda_i \langle \vec{u}_i \mid, \qquad (5.17)$$

and

$$(\mathbf{T}^R)_{0,R} = \sum_i \langle \vec{s}_0 \mid \vec{u}_i \rangle \lambda_i^R \langle \vec{u}_i \mid \vec{s}_R \rangle \qquad (5.18)$$

eqn (5.15) becomes

$$\sum_{\{s\}} s_0 s_R e^{-\beta \mathcal{H}_N} =$$

$$\sum_{s_0 s_R} \sum_{i,j} s_0 \langle \vec{s}_0 \mid \vec{u}_i \rangle \lambda_i^R \langle \vec{u}_i \mid \vec{s}_R \rangle s_R \langle \vec{s}_R \mid \vec{u}_j \rangle \lambda_j^{N-R} \langle \vec{u}_j \mid \vec{s}_0 \rangle. \qquad (5.19)$$

Moving the final matrix element to the beginning of the product and using eqn (5.16):

$$\sum_{\{s\}} s_0 s_R e^{-\beta \mathcal{H}_N} = \sum_{i,j} \langle \vec{u}_j \mid \mathbf{s}_0 \mid \vec{u}_i \rangle \lambda_i^R \langle \vec{u}_i \mid \mathbf{s}_R \mid \vec{u}_j \rangle \lambda_j^{N-R}. \qquad (5.20)$$

Hence, recalling the formula (5.8) for $\mathcal{Z}$,

$$\langle s_0 s_R \rangle_N = \frac{\sum_{i,j} \langle \vec{u}_j \mid \mathbf{s}_0 \mid \vec{u}_i \rangle \left(\frac{\lambda_i}{\lambda_0} \right)^R \langle \vec{u}_i \mid \mathbf{s}_R \mid \vec{u}_j \rangle \left(\frac{\lambda_j}{\lambda_0} \right)^{N-R}}{\sum_k \left(\frac{\lambda_k}{\lambda_0} \right)^N} \qquad (5.21)$$

where we have divided through by λ_0. It is then easy to see that in the thermodynamic limit only the terms in $j = 0$ and $k = 0$ survive

$$\langle s_0 s_R \rangle = \lim_{N \to \infty} \langle s_0 s_R \rangle_N = \sum_i \left(\frac{\lambda_i}{\lambda_0} \right)^R \langle \vec{u}_0 \mid \mathbf{s}_0 \mid \vec{u}_i \rangle \langle \vec{u}_i \mid \mathbf{s}_R \mid \vec{u}_0 \rangle \qquad (5.22)$$

$$= \langle \vec{u}_0 \mid \mathbf{s}_0 \mid \vec{u}_0 \rangle \langle \vec{u}_0 \mid \mathbf{s}_R \mid \vec{u}_0 \rangle + \sum_{i \neq 0} \left(\frac{\lambda_i}{\lambda_0} \right)^R \langle \vec{u}_0 \mid \mathbf{s}_0 \mid \vec{u}_i \rangle \langle \vec{u}_i \mid \mathbf{s}_R \mid \vec{u}_0 \rangle$$

$$= \langle s_0 \rangle \langle s_R \rangle + \sum_{i \neq 0} \left(\frac{\lambda_i}{\lambda_0} \right)^R \langle \vec{u}_0 \mid \mathbf{s}_0 \mid \vec{u}_i \rangle \langle \vec{u}_i \mid \mathbf{s}_R \mid \vec{u}_0 \rangle \qquad (5.23)$$

where in the final step we have used

$$\langle s_R \rangle = \langle \vec{u}_0 \mid \mathbf{s}_R \mid \vec{u}_0 \rangle \qquad (5.24)$$

which can be proved by a method entirely analogous to that followed above (see problem 5.1).

The correlation function (5.12) then follows immediately as

$$\Gamma_R = \sum_{i \neq 0} \left(\frac{\lambda_i}{\lambda_0}\right)^R \langle \vec{u}_0 \mid \mathbf{s}_0 \mid \vec{u}_i \rangle \langle \vec{u}_i \mid \mathbf{s}_R \mid \vec{u}_0 \rangle. \qquad (5.25)$$

Note that it depends on all the eigenvalues and eigenvectors of the transfer matrix. A much simpler formula is obtained for the correlation length (5.13). Taking the limit $R \to \infty$ the term $i = 1$ dominates the sum in eqn (5.25) and hence

$$\xi^{-1} = \lim_{R \to \infty} -\frac{1}{R} \ln \left\{ \left(\frac{\lambda_1}{\lambda_0}\right)^R \langle \vec{u}_0 \mid \mathbf{s}_0 \mid \vec{u}_1 \rangle \langle \vec{u}_1 \mid \mathbf{s}_R \mid \vec{u}_0 \rangle \right\} (5.26)$$

$$= -\ln(\lambda_1/\lambda_0). \qquad (5.27)$$

This formula has proved invaluable in work involving large transfer matrices—it is far easier numerically to find a small number of dominant eigenvalues than to completely diagonalize the matrix.

A point worth noting is that usually the Hamiltonian considered is translationally invariant. Hence the product of matrix elements in eqn (5.25) can be rewritten

$$\langle \vec{u}_0 \mid \mathbf{s}_0 \mid \vec{u}_i \rangle \langle \vec{u}_i \mid \mathbf{s}_R \mid \vec{u}_0 \rangle = \mid \langle \vec{u}_i \mid \mathbf{s}_0 \mid \vec{u}_0 \rangle \mid^2. \qquad (5.28)$$

We have also ignored the possibility that λ_i, $i \neq 0$, can be complex. This case is followed through in problem 5.3.

5.4 Results for the Ising model

Let us now return to the example considered in Section 5.1, the nearest neighbour Ising model in a magnetic field, to obtain explicit results for the quantities discussed in Sections 5.2 and 5.3. Diagonalizing the matrix (5.6) gives

$$\lambda_{0,1} = e^{\beta J} \cosh \beta H \pm \sqrt{e^{2\beta J} \sinh^2 \beta H + e^{-2\beta J}}, \qquad (5.29)$$

$$\langle \vec{u}_0 \mid = (\alpha_+, \alpha_-), \qquad \langle \vec{u}_1 \mid = (\alpha_-, -\alpha_+) \qquad (5.30)$$

where

$$\alpha_\pm^2 = \frac{1}{2} \left(1 \pm \frac{e^{\beta J} \sinh \beta H}{\sqrt{e^{2\beta J} \sinh^2 \beta H + e^{-2\beta J}}} \right). \qquad (5.31)$$

Using eqns (5.29)–(5.31) we shall write down expressions for the free energy per spin f, the magnetization per spin $\langle s \rangle$, the correlation function Γ, and the correlation length ξ, and check that they behave in a sensible way.

5.4.1   The free energy

From eqns (5.11) and (5.29)

$$f = -kT \ln \left\{ e^{\beta J} \cosh \beta H + \sqrt{e^{2\beta J} \sinh^2 \beta H + e^{-2\beta J}} \right\}. \quad (5.32)$$

As $\beta \to \infty$

$$f \to -kT \ln\{e^{\beta J}(\cosh \beta H + \sinh \beta H)\} = -J - H \quad (5.33)$$

which is the energy per spin as expected.

5.4.2   The magnetization

This can be obtained either by differentiating the negative of the free energy with respect to the magnetic field H, or by using eqn (5.24). One obtains

$$\langle s \rangle = (\alpha_+, \alpha_-) \begin{pmatrix} 1 & 0 \\ 0 & -1 \end{pmatrix} \begin{pmatrix} \alpha_+ \\ \alpha_- \end{pmatrix} \quad (5.34)$$

$$= \frac{e^{\beta J} \sinh \beta H}{\sqrt{e^{2\beta J} \sinh^2 \beta H + e^{-2\beta J}}}. \quad (5.35)$$

For non-interacting spins $J = 0$ (or equivalently $T = \infty$), this reduces to

$$\langle s \rangle = \tanh \beta H \quad (5.36)$$

as expected for a paramagnet. In zero field at any finite temperature $\langle s \rangle = 0$, as expected from the symmetry of the model, unless one takes the temperature to zero with H finite and then the field to zero

$$\lim_{H \to 0^\pm} \lim_{T \to 0} \langle s \rangle = \pm 1 \quad (5.37)$$

showing that there is a phase transition at zero temperature to a fully ordered ground state.

5.4.3   The correlation function

From eqn (5.25)

$$\Gamma_R = \left(\frac{\lambda_1}{\lambda_0}\right)^R \frac{e^{-2\beta J}}{e^{2\beta J} \sinh^2 \beta H + e^{-2\beta J}}. \quad (5.38)$$

For zero field this simplifies to

$$\Gamma_R(H = 0) = \tanh^R \beta J. \quad (5.39)$$

The zero-field correlation function is plotted as a function of R for different temperatures in Fig. 5.1. Note the expected decay with R for

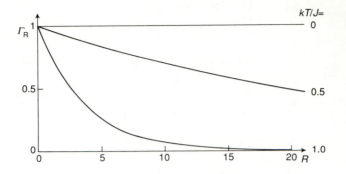

Fig.   5.1.   Dependence of the spin–spin correlation function of the one-dimensional Ising model in zero field on distance and temperature.

all $T \neq 0$. If the coupling is antiferromagnetic $(J < 0)$ the correlation function changes sign for odd R.

5.4.4   The correlation length

From eqn (5.27)

$$\xi^{-1} = -\ln \left\{ \frac{e^{\beta J} \cosh \beta H - \sqrt{e^{2\beta J} \sinh^2 \beta H + e^{-2\beta J}}}{e^{\beta J} \cosh \beta H + \sqrt{e^{2\beta J} \sinh^2 \beta H + e^{-2\beta J}}} \right\} . \qquad (5.40)$$

Check that as $T \to 0$, $\xi^{-1} \to 0$ signalling the expected phase transition and that as $T \to \infty$, $\xi^{-1} \to \infty$.

5.5   Problems

5.1 Prove that, in the thermodynamic limit, the average value of the spin $\langle s \rangle$ is given by

$$\langle s \rangle = \langle \vec{u}_0 \mid \mathbf{s} \mid \vec{u}_0 \rangle$$

where $\mathbf{s}$ is the diagonal matrix with eigenvalues equal to the possible values of the spin and $\mid \vec{u}_0 \rangle$ is the eigenvector corresponding to the largest eigenvalue of the transfer matrix.

5.2 (i) Write down the transfer matrix for the one-dimensional, q-state Potts model which is described by the Hamiltonian

$$\mathcal{H} = -J \sum_i \delta_{\sigma_i \sigma_{i+1}}, \qquad \sigma_i = 1, 2 \ldots q.$$

(ii) Show that the largest eigenvalue is $e^{\beta J} + q - 1$ and that the remaining eigenvalues are all degenerate and take the value $e^{\beta J} - 1$.

(iii) Write down expressions for the free energy and correlation length of the model and show that they take sensible values in the limits of zero and infinite temperature.

5.3 Write down the transfer matrix for the one-dimensional spin-1 Ising model in zero field which is described by the Hamiltonian

$$\mathcal{H} = -J \sum_i s_i s_{i+1}, \qquad s_i = \pm 1, 0.$$

Hence calculate the free energy per spin of this model and show that it has the expected behaviour in the limits $T \to 0$ and $T \to \infty$.

[Answer: $f = -kT \ln\{(1+2\cosh\beta J+[(2\cosh\beta J-1)^2+8]^{1/2})/2\}.$]

5.4 Consider a transfer matrix with largest eigenvalue λ_0 whose eigenvalues of second largest modulus form a complex conjugate pair $|\lambda_1| e^{\pm i\theta}$. Prove that the correlation length is given by

$$\xi^{-1} = -\ln(|\lambda_1|/\lambda_0)$$

and that the correlations decay with a wavevector θ.

5.5 The one-dimensional, three-state chiral clock model is described by the Hamiltonian

$$\mathcal{H} = -J \sum_i \cos\{2\pi(n_i - n_{i+1} + \Delta)/3\}, \qquad n_i = 0, 1, 2.$$

(i) Write down the transfer matrix and show that its eigenvalues and eigenvectors are

$$\lambda_0 = a + b + c \qquad |\vec{u}_0\rangle = \frac{1}{\sqrt{3}}(1,1,1)$$

$$\lambda_1 = a + \omega b + \omega^2 c \qquad |\vec{u}_1\rangle = \frac{1}{\sqrt{3}}(1,\omega,\omega^2)$$

$$\lambda_2 = \lambda_1^* = a + \omega^2 b + \omega c \qquad | \, \bar{u}_2 \rangle = \frac{1}{\sqrt{3}}(1, \omega^2, \omega)$$

where ω is a complex cube root of unity and

$$a = e^{\beta J \cos\{2\pi\Delta/3\}}, \qquad b = e^{\beta J \cos\{2\pi(\Delta - 1)/3\}},$$
$$c = e^{\beta J \cos\{2\pi(\Delta + 1)/3\}}.$$

(ii) Hence determine the free energy f, correlation function Γ_R, correlation length ξ, and wavevector associated with the decay of correlations θ.

(iii) Comment on the limit $\Delta \to 0$.

[Answers: $f = -kT \ln(a + b + c)$;

$\Gamma_R = \frac{2}{3} \left(\frac{|\lambda_1|}{\lambda_0} \right)^R \cos R\theta$;

$\xi^{-1} = \ln\{(a + b + c)/ \, | \, a + \omega^2 b + \omega c \, |\}$;

$\theta = \tan^{-1}\{\sqrt{3}(b - c)/(2a - b - c)\}$.]

5.6 Show that the transfer matrix for the spin-1/2 Ising model with first and second neighbour interactions

$$\mathcal{H} = -J_1 \sum_i s_i s_{i+1} - J_2 \sum_i s_i s_{i+2}, \qquad s_i = \pm 1$$

may be written in terms of $x = e^{\beta J_1}$ and $y = e^{\beta J_2}$ as

(s_i, s_{i+1})	(s_{i+2}, s_{i+3})			
	$1,1$	$1,-1$	$-1,1$	$-1,-1$
$1,1$	$x^2 y^2$	x^2	1	y^{-2}
$1,-1$	x^{-2}	$x^{-2}y^2$	y^{-2}	1
$-1,1$	1	y^{-2}	$x^{-2}y^2$	x^{-2}
$-1,-1$	y^{-2}	1	x^2	$x^2 y^2$

if two spins are added by each transfer matrix or

(s_i, s_{i+1})	(s_{i+1}, s_{i+2})			
	$1,1$	$1,-1$	$-1,1$	$-1,-1$
$1,1$	xy	xy^{-1}	0	0
$1,-1$	0	0	$x^{-1}y$	$x^{-1}y^{-1}$
$-1,1$	$x^{-1}y^{-1}$	$x^{-1}y$	0	0
$-1,-1$	0	0	xy^{-1}	xy

if a single spin is added at each step.

An analysis of this model, which circumvents diagonalizing the 4×4 transfer matrix is given in Stephenson, J. (1970). *Canadian Journal of Physics*, **48**, 1724.

5.7 A simple model of an interface is the solid-on-solid model illustrated in Fig. 5.2. In each column of the lattice, i, the interface lies at a position n_i which is constrained to be single-valued. Thus overhangs and excitations of the bulk are forbidden. A solid-on-solid Hamiltonian which allows description of the binding of the interface to a substrate at $n_i = 0$ is

$$\mathcal{H} = J \sum_i |\, n_i - n_{i+1}\,| - K \sum_i \delta_{n_i,0}; \quad n_i = 0, 1, 2 \ldots .$$

(i) Write down the transfer matrix of this model in terms of

$$w = e^{-J/kT}, \quad \kappa = e^{K/kT}.$$

(ii) By considering an eigenvector of the form

$$(\psi_0, \ \cos(q + \theta), \ \cos(2q + \theta) \ldots)$$

show that there is a continuous spectrum of eigenvalues

$$(1 - w)/(1 + w) \le \lambda \le (1 + w)/(1 - w).$$

(iii) Show that, for $\kappa > (1 - w)^{-1}$, there is also a bound state eigenvector of the form

$$(\psi_0, \ e^{-\mu}, \ e^{-2\mu} \ldots)$$

which corresponds to an eigenvalue

$$\lambda_0 = \frac{\kappa(1 - w^2)(\kappa - 1)}{\kappa(1 - w^2) - 1}.$$

(iv) Show that, where it exists, λ_0 is the largest eigenvalue. This means that it dominates the thermodynamics and the interface binds to the substrate at $\kappa_c = (1 - w)^{-1}$. What is the eigenvector corresponding to the largest eigenvalue at this point?

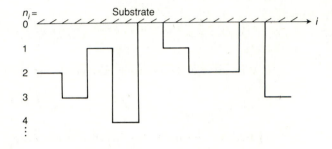

Fig. 5.2. The solid-on-solid model of an interface on a semi-infinite lattice. In each column of the lattice, i, the interface lies at a position $n_i \geq 0$.

6

Series expansions

Exact power series expansions for thermodynamic functions have in the past proved an invaluable aid to understanding the critical behaviour of insoluble models. Indeed, the first suggestions of power law singularities at criticality were based on such analyses. Immediately before the renormalization group was proposed, work using series expansions had led to a large body of evidence that exponents had universal properties, that they were the same above and below the critical temperature, and that mean-field values set in abruptly above four dimensions.

'One somewhat intriguing result that has arisen from the [series] analysis of lattices with $d > 3$ is the following: rather than the anticipated mean-field behaviour setting in gradually as $d \to \infty$ the mean-field critical point exponents appear to be obtained for all values of $d \geq 4$.'[1]

Series expansions remain, in many cases, one of the most accurate ways of estimating critical exponents. The idea is to find a systematic way of calculating classes of contributions to the partition function which can be obtained exactly and hope that the successive approximations can be extrapolated to give information about critical properties.

Two expansion procedures will be considered in this chapter. The first is high temperature series where the Boltzmann factor is expanded in powers of the inverse temperature and the trace taken term by term. In the second, low temperature expansions, configurations are counted in order of their importance as the temperature is increased from zero: starting from the ground state the series is constructed by successively adding terms from $1, 2, 3, \ldots$ flipped spins.

As the order of the expansion is increased the number and complexity of contributing terms also increases rapidly. A rule of thumb is

---

[1]Stanley, H. E. (1971). *Introduction to phase transitions and critical phenomena*, Ch. 4. (Oxford University Press, Oxford).

that the work involved in calculating the last term is the same as that needed to calculate all the preceding terms. As we shall see, each term in the series can be represented by graphs on a lattice and constructing the series comes down to counting the allowed graphs, which is usually done using a computer.

The expansions obtained can be used to give an approximation to the thermodynamic properties of a model at low and high temperatures. However, they can be, and more often are, used to study critical properties. The hope is that the expansions are sufficiently well behaved that information about their singularities can be obtained from the limited number of terms available.

The radius of convergence of a series is determined by the singularity which lies nearest to the origin in the complex plane. If this is on the real axis it can, in general, be identified as the critical point whose value and associated exponents can be estimated. Even if the leading singularity lies in the complex plane and is non-physical there are still analysis techniques which can be used to extract the critical behaviour.

There is no rigorous justification that the series expansions are convergent. However, it is widely believed that this procedure works, and not only works but works well—at present the argument is over the third decimal place in series estimates of the exponents of the three-dimensional Ising model. Confidence in the method lies in the large body of circumstantial evidence available. Series expansions agree well with high accuracy Monte Carlo simulations, with renormalization group results, and with exact results for soluble models where these are available. Comparable results are obtained from the analysis of series for different thermodynamic variables and from low and high temperature expansions. Moreover, usually the series behave in a sensible way; as extra terms are added the extrapolated results converge stably. More recently the understanding of the universality of critical exponents has provided another benchmark; the scatter of results for different lattice types provides some estimate of the error bars in the expansion results. However, historically, it was the results from series expansions that suggested universality.

6.1   High temperature series expansions

We first consider the high temperature series expansion for the two-dimensional, zero-field Ising model, defined by the Hamiltonian (3.1) with $H = 0$, on a square lattice. Although this has many simplifying features it illustrates the important ideas involved in the construction of series expansions. Because, for the Ising model, $s_i s_j = \pm 1$, we may write

$$e^{\beta J s_i s_j} = \cosh \beta J + s_i s_j \sinh \beta J \equiv \cosh \beta J (1 + s_i s_j v). \quad (6.1)$$

$v = \tanh \beta J$ is the natural high temperature expansion variable for this problem: $v \to 0$ as $T \to \infty$ as required.

Using eqn (6.1) to rewrite the partition function leads to a form that can be easily expanded in powers of v

$$\mathcal{Z} = \sum_{\{s\}} \prod_{\langle ij \rangle} e^{\beta J s_i s_j} \quad (6.2)$$

$$= (\cosh \beta J)^{\mathcal{B}} \sum_{\{s\}} \prod_{\langle ij \rangle} (1 + s_i s_j v) \quad (6.3)$$

$$= (\cosh \beta J)^{\mathcal{B}} \sum_{\{s\}} (1 + v \sum_{\langle ij \rangle} s_i s_j$$

$$+ v^2 \sum_{\langle ij \rangle ; \langle kl \rangle} s_i s_j s_k s_l + ...) \quad (6.4)$$

where $\mathcal{B}$ is the number of bonds on the lattice. The aim is to count the number of contributions to $\mathcal{Z}$ which are of order v^n up to as large values of n as possible. The easiest way is to use the correspondence between the terms in eqn (6.4) and graphs on the square lattice. Each product of a pair of spins, $s_i s_j$, can be associated with the bond on the lattice which joins sites i and j. Each term of order v can be represented by a single bond. Terms of order v^2 correspond to two bonds which may, or may not, touch and so on. Therefore each term of order v^n is in one-to-one correspondence with a graph with n edges on the square lattice. Examples are shown in Fig. 6.1.

We have to consider not only the number of graphs at a given order but also their contribution to the partition function. Fortunately this is zero in many cases. Because $s_i = \pm 1$

$$\sum_{\{s\}} (s_i^{n_i} s_j^{n_j} s_k^{n_k} ...) = 2^N \quad \text{(all } n_i \text{ even)}$$
$$= 0 \quad \text{(otherwise)} \quad (6.5)$$

where N is the number of spins on the lattice. Hence only products in which every spin operator appears an even number of times contribute. Graphically these terms correspond to closed loops; no free ends are allowed. Each contributes the same weight, 2^N.

So finding the contribution to the partition function of order n is reduced to the problem of counting the number of closed loops of n

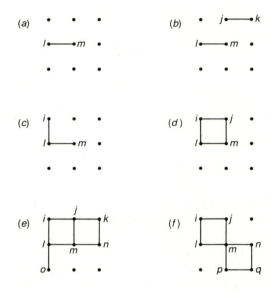

Fig. 6.1. Graphs on the square lattice, each of which corresponds to a product of spins in the sum in eqn (6.4): (a) $s_l s_m$, (b) $s_l s_m s_j s_k$, (c) $s_i s_l^2 s_m$, (d) $s_i^2 s_j^2 s_l^2 s_m^2$, (e) $s_i^2 s_j^3 s_k^2 s_l^3 s_m^3 s_n^2 s_o$, ($f$) $s_i^2 s_j^2 s_l^2 s_m^4 s_n^2 s_p^2 s_q^2$. Only ($d$) and ($f$), where the number of bonds at each vertex is even, give a non-zero contribution to the partition function.

bonds that can be put on the square lattice. Remember that every position and orientation of the loops will give a contribution to the partition function. Terms to order v^{10} are shown in Table 6.1. Reading from the table and using eqn (6.4) gives the leading terms in the high temperature series expansion for the partition function of the two-dimensional Ising model on the square lattice.

$$\mathcal{Z} = (\cosh \beta J)^{\mathcal{B}} 2^N \{1 + Nv^4 + 2Nv^6$$
$$+ \frac{1}{2} N(N+9)v^8 + 2N(N+6)v^{10} + O(v^{12})\}. \quad (6.6)$$

The free energy follows as usual from the logarithm of the partition function. Taking the logarithm of eqn (6.6), noting for the square lattice that $\mathcal{B} = 2N$, and expanding for small v gives

$$\mathcal{F} = -NkT\{\ln 2 + v^2 + \frac{3}{2}v^4 + \frac{7}{3}v^6 + \frac{19}{4}v^8 + \frac{61}{5}v^{10} + O(v^{12})\}. \quad (6.7)$$

Happily terms with counts proportional to N^2 have dropped out. It can be proved that only terms proportional to N survive at all orders of the expansion, as must be the case if the free energy is to be extensive. This is an example of the linked cluster theorem[2]. It means that it is often possible to formulate rules for calculating the contribution from disconnected diagrams, which are those responsible for terms non-linear in N, without having to obtain the full count.

Having written down an expansion for the free energy the specific heat series can be obtained by differentiation. If a magnetic field is included in the original Hamiltonian—which leads to a relaxation of the constraint on even vertices—the susceptibility series can be generated. This case is considered in problem 6.7. It is also possible to write down series expansions for the correlation functions (see problem 6.6).

How far is it possible to get? The two-dimensional Ising model on a square lattice can be solved exactly in zero field and therefore the series expansion, should one be interested, can be written down to all orders in v. The greatest interest lies in three dimensions where there is continuing progress in refining values for the critical exponents for comparison with increasingly accurate experiments. At the time of writing the susceptibility series for the Ising model on the body-centred cubic lattice is complete to order v^{21}, the specific heat series to order

---

[2]Wortis, M. (1974). *Linked cluster expansion*. In *Phase transitions and critical phenomena*, Vol 3 (eds C. Domb and M. S. Green), p.113. (Academic Press, London).

Table 6.1.   The configurations, together with their counts, which
contribute to the high temperature expansion of the partition function
of the Ising model on a square lattice

Order	Contributing graphs	Count
v^4		N
v^6		$2N$
v^8		$N(N-5)/2$
		$4N$
		N
		$2N$
v^{10}		$2N(N-8)$
		$2N$
		$8N$
		$4N$
		$8N$
		$4N$
		$2N$

v^{14}. The argument is about the third decimal place in the values of the critical exponents, the fourth in the value of the critical temperature.

The identity (6.1) which is a property of the spin-1/2 Ising model introduces simplifications helpful to both practitioner and pedagogue. More generally the expansion of the partition function is written

$$\mathcal{Z} = \sum_{\{s\}} e^{-\beta \mathcal{H}} = \sum_{\{s\}} (1 - \beta \mathcal{H} + \beta^2 \mathcal{H}^2/2! - \dots) \qquad (6.8)$$

and the problem is to evaluate the trace of powers of the Hamiltonian. These are just traces of products of spins which it is helpful to identify with graphs on a lattice as before. However, in general, multiple bonds are allowed and the weights depend on the topology of the graphs. Rules pertinent to a given model are drawn up and many ingenious ways of doing the counting which lead to efficient numerical algorithms have been documented in the literature[3].

These are details best left to the expert, but it is important to point out that high temperature series have been applied widely to Ising models of all spin magnitudes and with further-neighbour and long-range interactions, other discrete models, such as Potts models, and continuous spin systems. The technique has also proved useful in geometrical problems such as percolation, self-avoiding walks, and in studying the field theories used in particle physics.

6.2 Low temperature series expansions

High temperature expansions cannot give any information about properties below the critical temperature. Therefore, to obtain a complete picture, low temperature expansions are also needed. At low temperatures for models with discrete spin variables[4] the dominant contribution to the partition function is from states where few spins are flipped relative to their value in the ground state. To exploit this we choose to order the terms in the partition function sum

$$\mathcal{Z} = e^{-E_0/kT} (1 + \sum_{n=1}^{\infty} \Delta \mathcal{Z}_N^{(n)}) \qquad (6.9)$$

---

[3]Domb, C. and Green, M. S. (eds) (1974). *Phase transitions and critical phenomena*, Vol 3. (Academic Press, London).

[4]For Heisenberg models there are no excitations involving discrete energy steps and spin-wave theory is appropriate.

where $\Delta \mathcal{Z}_N^{(n)}$ is the sum of Boltzmann factors (with energy, for convenience, being measured relative to the ground state energy E_0) of all states where n spins are flipped relative to the ground state.

For the Ising model each wrong bond associated with flipping a spin has an energy, relative to the ground state, of $2J$ and hence a Boltzmann weight

$$x = e^{-2J/kT}. \qquad (6.10)$$

x is the natural expansion variable for the low temperature series. For a single spin flip, which in two dimensions generates four dissatisfied bonds, the Boltzmann weight is x^4; for two spin flips it is x^8 unless they are nearest neighbours, in which case only six wrong bonds are generated giving a Boltzmann weight x^6.

Two factors go to make up the $\mathcal{Z}_N^{(n)}$; the number of ways of flipping n spins with given Boltzmann weights, or counts, and the corresponding Boltzmann weights themselves. These are listed in Table 6.2; once the counts are sorted out the Boltzmann weights follow easily. Adding the terms in the table gives the leading behaviour of the low temperature expansion of the partition function of the two-dimensional Ising model:

$$\begin{aligned} \mathcal{Z} &= e^{-E_0/kT}\{1 + Nx^4 + 2Nx^6 + \frac{1}{2}N(N+9)x^8 \\ &\quad + 2N(N+6)x^{10} + O(x^{12})\}. \end{aligned} \qquad (6.11)$$

Note that there is not a one-to-one correspondence between the number of spin flips and the powers of x appearing in the Boltzmann weights. Four-flip terms contribute at orders between x^8 and x^{16}. The expansion is in powers of x, not in the number of spin flips.

6.3   The one-dimensional Ising model

Because the ordering temperature of the one-dimensional Ising model is zero the high temperature series expansion is expected to be convergent at all finite temperatures. It can be written down exactly and easily. For a lattice with free boundaries and N spins there are no closed graphs. Hence, from eqn (6.4),

$$\mathcal{Z} = 2^N \cosh^{N-1} \beta J \qquad (6.12)$$

where the powers of 2 come from taking the trace of unity and $\mathcal{B} = N - 1$. For periodic boundary conditions and N spins the graph where all bonds are occupied is allowed and

Table 6.2. The configurations, together with their counts and Boltzmann weights, which contribute to the low temperature expansion of
the partition function of the two-dimensional Ising model on a square
lattice to order x^{10}

Number of flipped spins	Configuration	Count	Boltzmann weight
1		N	x^4
2		$2N$	x^6
		$N(N-5)/2$	x^8
3		$2N$	x^8
		$4N$	x^8
		$2N(N-8)$	x^{10}
		$N(N^2-15N+62)/6$	x^{12}
4		N	x^8
		$8N$	x^{10}
		$2N$	x^{10}
		$4N$	x^{10}
		$4N$	x^{10}

(terms up to x^{16})

| 5 | | $8N$ | x^{10} |

(terms up to x^{20})

| 6 | | $2N$ | x^{10} |

(terms up to x^{24})

Table 6.3. The configurations, together with their counts and Boltzmann weights, which contribute to the low temperature expansion of
the partition function of the one-dimensional Ising model

Number of flipped spins	Configuration	Count	Boltzmann weight
1	●	N	x^4
2	●—●	N	x^2
	● ●	$N(N-3)/2$	x^4
3	●—●—●	N	x^2
	●—● ●	$N(N-4)$	x^4
	● ● ●	$N(N^2 - 9N + 20)/6$	x^6
4	●—●—●—●	N	x^2

$$\vdots$$

(terms up to x^8)

$$\mathcal{Z} = 2^N \cosh^N \beta J (1 + v^N).\tag{6.13}$$

The first few terms in the low temperature expansion of the one-
dimensional Ising model are shown in Table 6.3. Flipping any number
of neighbouring spins gives the same Boltzmann weight, x^2. So the
series diverges at $x = 0$ as expected for a model with a zero temperature
phase transition.

6.4   Analysis of series expansions

Summing the terms in a series expansion can give an approximation to
the low or high temperature behaviour of a given spin model. However,
historically there has been far more interest in using the expansions to
predict the value of the critical temperature and the associated critical
exponents. To do this the singular behaviour must be extracted from
a regular expansion.

The radius of convergence of a power series is determined by the singularity nearest the origin in the complex plane. If this fortuitously lies on the positive real axis it can be identified with the critical temperature and a simple analysis of successive coefficients allows the scaling behaviour to be extracted.

As $T \to T_c$ a thermodynamic function $Y(t)$ is expected to obey the scaling form

$$Y(t) \sim t^{-\lambda} \cdot \qquad (6.14)$$

where t is the reduced temperature defined by eqn (2.18). Writing eqn (6.14) in terms of a typical high temperature expansion variable $y = \beta J$ and expanding in y gives

$$\tilde{Y}(y) \sim \left(\frac{y}{y_c}\right)^{\lambda} [1 + \frac{\lambda y}{y_c} + \frac{\lambda(\lambda+1)}{2!}\left(\frac{y}{y_c}\right)^2 + \dots$$

$$\dots + \frac{\lambda(\lambda+1)\dots(\lambda+(n-1))}{n!}\left(\frac{y}{y_c}\right)^n + \dots]$$

$$\equiv \sum_n a_n y^{n+\lambda} \qquad (6.15)$$

where $y_c = \beta_c J$. Comparing the coefficients of $y^{n+\lambda}$ and $y^{n+\lambda+1}$ in the expansion one obtains the simple result

$$\frac{a_n}{a_{n-1}} = \frac{1}{y_c} + \frac{(\lambda-1)}{y_c n}. \qquad (6.16)$$

Corrections to scaling will lead to deviations from eqn (6.16) for all finite n. However, the hope is that a plot of a_n/a_{n-1} versus $1/n$ will give an intercept and slope approximating to y_c^{-1} and $(\lambda-1)y_c^{-1}$ respectively.

An example is shown in Fig. 6.2 for the reduced susceptibility (that is, the susceptibility divided by its value in the non-interacting limit) series of the spin-1/2 Ising model on lattices of different dimensionalities. These series are well behaved and, even for the low orders shown, converge rather smoothly to the asymptotic behaviour described by eqn (6.16).

Life becomes more complicated if the closest singularity to the origin does not lie on the real axis. The signal of this is that the ratio of successive coefficients does not converge smoothly. In this case a common approach is to calculate the series for the logarithmic derivative of a thermodynamic function

$$\frac{d}{dT}\{\ln Y(t)\} \sim -\frac{\lambda}{T - T_c} \qquad (6.17)$$

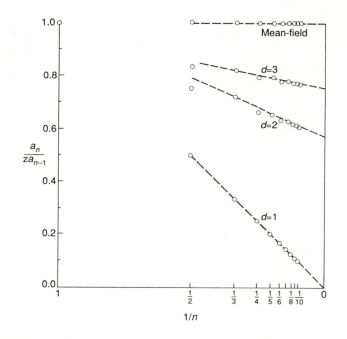

Fig. 6.2. The ratio of successive coefficients, a_n/za_{n-1}, of the reduced susceptibility series of the spin-1/2 Ising model on lattices of different dimensionality plotted against $1/n$. For ease of display the data are normalised by the coordination number of the lattice z. The expected limiting behaviour, given by eqn (6.16), is shown by the dotted lines with the parameters in the equation taken from the exact or best series results available. After Stanley, H. E. (1971). *Introduction to phase transitions and critical phenomena.* (By permission of Oxford University Press, Oxford).

Table 6.4. Estimates of the critical point v_c and the critical exponent γ (in brackets) for the Ising model on a square lattice from the poles and residues of the [L,M] Padé approximants to the series for the logarithmic derivative of the susceptibility. After Gaunt, D. S. and Guttmann, A. J. (1974). *Asymptotic analysis of coefficients.* In *Phase transitions and critical phenomena*, Vol 3 (eds C. Domb and M. S. Green), p.181. (Academic Press, London)

M	$L = M - 1$		$L = M$	
1	0.50000	(-2.0000)	0.28571	(-0.6531)
2	0.38871	(-1.4017)	0.41119	(-1.6546)
3	0.40888	(-1.6186)	0.40927	(-1.6257)
4	0.40877^a	(-1.6171)	0.41645	(-1.7974)
5	0.41019	(-1.6383)	0.41217^a	(-1.6823)
6	0.41484	(-1.7782)	0.41413	(-1.7458)
7	0.414249	(-1.7515)	0.414211	(-1.7496)
8	0.414214	(-1.7498)	0.414213	(-1.7498)
9	0.414213	(-1.7498)	0.414214^a	(-1.7498)
10	0.414202	(-1.7484)	0.414213	(-1.7497)
	exact values		$\sqrt{2} - 1$	$-7/4$

[a] approximant with an intervening spurious pole

which has a simple pole at $T = T_c$ with residue $-\lambda$. An $[L, M]$ Padé approximant

$$\frac{P_L(y)}{Q_M(y)} = \frac{p_0 + p_1 y + p_2 y^2 + \ldots + p_L y^L}{1 + q_1 y + q_2 y^2 + \ldots + q_M y^M} \tag{6.18}$$

is then constructed with the $L + M + 1$ coefficients, p_i, q_i, chosen so that the expansion of the Padé agrees with the first $(L + M + 1)$ terms in the series expansion. The hope is that the denominator of the Padé will reproduce the pole at T_c.

For a series with $n_0 + 1$ terms Padé approximants can be written down for all L, M such that $L + M \le n_0$. $L \approx M$ usually gives the best results. A comparison of the values which result from different approximants is used to give some feel for the stability of the procedure[5]. An example is given in Table 6.4.

6.5 Problems

6.1 Consider an interface in a one-dimensional Ising model,

$$s_i = -1, \quad i < 0; \qquad s_i = 1, \quad i \ge 0.$$

By writing down the energy and entropy associated with such an excitation argue that the one-dimensional Ising model cannot sustain long-range order for any non-zero temperature.

6.2 Show that there is a one-to-one correspondence between the terms in the high and low temperature expansions of the spin-1/2, zero-field Ising model on the square lattice. This model is said to be self-dual.

Hence argue that, if the critical temperature, T_c, is unique, it must be given by

$$e^{-2J/kT_c} = \tanh J/kT_c$$

or

$$J/kT_c = \frac{1}{2} \ln(1 + \sqrt{2}).$$

---

[5]Relatively little is known about the way in which Padé approximants converge, but see the reference given in Table 6.4 for a summary of the results available.

6.3 The exact result for the spontaneous magnetization per spin of the spin-1/2 Ising model on the square lattice is

$$\langle s \rangle = (1+u)^{1/4}(1-u)^{-1/2}(1-6u+u^2)^{1/8}, \quad u = e^{-4J/kT}. \quad (6.19)$$

Expanding this result gives

$$\langle s \rangle = 1 - 2u^2 - 8u^3 - 34u^4 - 152u^5 - 714u^6 - 3472u^7 - \dots \quad (6.20)$$

(i) Generalize the low temperature expansion for this model, given in Section 6.2, to include a non-zero field, H. Hence obtain the series for the zero-field magnetization to terms $O(u^5)$. Check that your answer agrees with the exact result.

(ii) Use the ratio method to obtain an estimate for the critical temperature and the exponent β from the expansion (6.20). Compare with the exact results, $u_c = 3 - 2\sqrt{2}$, $\beta = 1/8$, which follow immediately from eqn (6.19).

6.4 For the three-dimensional spin-1/2 Ising model on a cubic lattice the low temperature expansion for the partition function is

$$\mathcal{Z}_N = e^{-\beta N E_0}\{1 + Nx^6 + 3Nx^{10} + \tfrac{1}{2}N(N-7)x^{12}$$
$$+15Nx^{14} + 3N(N-11)x^{16} + O(x^{18})\}$$

where N is the number of spins on the lattice and E_0 is the ground state energy per spin. List the graphs, and the associated counts and Boltzmann weights, that contribute to this expression. Comment on the order of the correction term.

6.5 In performing the low temperature expansion for the q-state Potts model

$$\mathcal{H} = -J \sum_{\langle ij \rangle} \sigma_i \sigma_j, \quad \sigma_i = 1, 2 \dots q$$

account must be taken of configurations in which the spin flips to each of the $(q-1)$ other states. Bearing this in mind list the Boltzmann weights that would be associated with the configurations listed in Table 6.2 for the q-state Potts model.

6.6 Use a high temperature series expansion to show that the two-spin correlation function of the one-dimensional Ising model in zero field is

$$\Gamma_R = \tanh^R \beta J.$$

6.7 This problem deals with the extension of the high temperature series expansion of the spin-1/2 Ising model, developed in Section 6.1, to include a field term.

(i) Show that the partition function of the spin-1/2 Ising model in a field can be written

$$\mathcal{Z} = \cosh^{\mathcal{B}} \beta J \cosh^{N} \beta H \sum_{\{s\}} \prod_{\langle ij \rangle} (1 + s_i s_j v) \prod_{i} (1 + s_i y)$$

where $v = \tanh \beta J$; $y = \tanh \beta H$, and $\mathcal{B}$ and N are the number of bonds and sites on the lattice respectively.

(ii) Show that the terms in $\mathcal{Z}$ can be represented by graphs on a lattice where each graph with l bonds and m odd vertices contributes a factor $2^{N} v^{l} y^{m}$, and hence that $\mathcal{Z}$ can be rewritten

$$\mathcal{Z} = \cosh^{\mathcal{B}} \beta J (2 \cosh \beta H)^{N} (1 + \mathcal{S}_0 + y^2 \mathcal{S}_2 + y^4 \mathcal{S}_4 + \ldots) \quad (6.21)$$

where $\mathcal{S}_m(v, N)$ is the contribution from all graphs with m odd vertices.

Taking the logarithm of eqn (6.21) and using the Linked Cluster Theorem[6] gives

$$-\beta \mathcal{F} = \mathcal{B} \ln \cosh \beta J + N \ln 2 \cosh \beta H + (\mathcal{S}_0' + y^2 \mathcal{S}_2' + y^4 \mathcal{S}_4' + \ldots) \quad (6.22)$$

where the prime denotes that, for a given graph, only the part of the count proportional to N must be included.

(iii) Differentiate eqn (6.22) to show that the zero-field susceptibility is

$$\chi = \beta N + 2\beta \mathcal{S}_2'.$$

(iv) List the low order graphs with two odd vertices to show that the first few terms in the high temperature expansion of the zero-field susceptibility of the spin-1/2 Ising model on the square lattice are

$$\chi = \beta N + 2\beta (2v + 6v^2 + 18v^3 + \ldots).$$

---

[6]See eqn (6.7) for an explicit example of how this works.

7

Monte Carlo simulations

It could be argued that current physics research can be divided into three areas—theoretical, experimental, and computational. Numerical approaches, in which systems are mimicked as accurately as possible using a computer or in which computer models are set up to provide well-behaved experimental systems are increasingly providing a bridge between theory and experiment. The limitations on what can be done are set by the computational resources available.

A powerful numerical approach is the Monte Carlo method. It was introduced in 1953 at the dawn of the computer age and its range of applicability and accuracy have continued to increase with the development of more advanced computer technology. One of the simplest and most natural applications, which we shall focus on here, is to discrete spin models. However the technique is very widely used: to study continuous spin systems, fluids, polymers, disordered materials, and lattice gauge theories. Some examples are given at the end of this chapter.

7.1 Importance sampling

A common aim in statistical mechanics is to find the value of a thermodynamic variable, such as the energy or the magnetization, which is a weighted sum over all states in phase space

$$\langle A \rangle = \frac{\sum_{\{s\}} A e^{-\beta \mathcal{H}}}{\sum_{\{s\}} e^{-\beta \mathcal{H}}}. \tag{7.1}$$

For an Ising model on a lattice of N sites the sum is over 2^N configurations. This is a number which increases very quickly with N and a direct evaluation is feasible only for $N \lesssim 40$.

The first way one might try to get round this is to choose randomly a sample of the spin configurations, $\{s\}$, and, weighing them appropriately according to eqn (7.1), work out an estimate of the required average. This approach may be familiar as it is a standard technique used for the evaluation of integrals. However, it fails here because of the rapid variation of the Boltzmann factor, $e^{-\beta\mathcal{H}}$, with energy. Very few of the chosen configurations will be weighted by a sufficiently large factor to make a significant contribution to the average and a very unreliable estimate will result.

This problem occurs because only an extremely restricted part of configuration space is important in determining the averages. This we already know from statistical mechanics—the system spends the vast majority of its time in states with thermodynamic parameters within $O(1/\sqrt{N})$ of those describing thermodynamic equilibrium. Therefore it would seem sensible to restrict the sampling to these states. This is a technique known as importance sampling. But how to generate such a set of states? To try to find the probability distribution exactly the partition function would need to be calculated and this is tantamount to going back to the original problem of summing over an impossibly large number of states.

Luckily it turns out to be possible to generate a Markov chain of configurations (a sequence of states each of which depends only on the preceding one) which has the property that $\hat{A}_n$, the average of A over n successive states, converges to the thermodynamic average defined in eqn (7.1)

$$\hat{A}_n = \langle A \rangle + O(n^{-1/2}). \tag{7.2}$$

In the limit $n \to \infty$ each state is weighted by its Boltzmann factor, $e^{-\beta E}$. The disadvantage of this approach is that successive states of the Markov chain are highly correlated, which means that a much longer sequence of sample configurations is needed to achieve a given accuracy than if this were not the case.

The conditions on the transition probability between Markov states needed to achieve the result (7.2) are physically transparent. The transition probability must be normalized. It must be ergodic, that is all states must eventually be accessible. Finally, a sufficient condition is that it must obey detailed balance[1].

This does not specify the transition probability uniquely. The choice often used in Monte Carlo simulations is the Metropolis algorithm. A final state, $\{s\}_f$, is chosen from an initial state, $\{s\}_i$, by

---

[1]Parisi, G. (1988). *Statistical field theory*, p.346. (Addison-Wesley, Wokingham).

flipping one or more spins. The probability that the system is allowed to move from i to f is

$$
\begin{aligned}
P(\{s\}_i \to \{s\}_f) &= e^{-\beta(E_f - E_i)} & \text{if } E_f > E_i \\
&= 1 & \text{if } E_i \leq E_f
\end{aligned}
\tag{7.3}
$$

where E_i and E_f are the energies of the initial and final states respectively.

There is a physically intuitive argument that shows that with this choice of transition probabilities the system tends asymptotically ($n \to \infty$) to a steady state in which the probability of a given configuration is $e^{-\beta E\{s\}}$. Consider m_r systems in a state $\{s\}_r$ and m_t in a state $\{s\}_t$ such that $E_t < E_r$. Using random numbers it is possible to construct a move such that the *a priori* probability of moving from state r to t is the same as that to move from t to r. (This is feasible but not always the case in realistic simulations.) Then, using eqns (7.3), the number of transitions from r to t and from t to r are

$$
M_{r \to t} \propto m_r
\tag{7.4}
$$

$$
M_{t \to r} \propto m_t e^{-\beta(E_r - E_t)}
\tag{7.5}
$$

respectively. The net number of transitions is

$$
\Delta M_{r \to t} \propto \{m_r - m_t e^{-\beta(E_r - E_t)}\}.
\tag{7.6}
$$

The system will converge to a steady state where $\Delta M_{r \to t} = 0$ or

$$
\frac{m_r}{m_t} = \frac{e^{-\beta E_r}}{e^{-\beta E_t}}.
\tag{7.7}
$$

7.2   Practical details

The steps involved in setting up a Monte Carlo simulation for a simple spin model are listed in the flow chart in Table 7.1. This is the basis of the program used to generate the spin configurations in Fig. 1.8. The procedure can be thought of in three parts. We concentrate in this section on the details of how to set up the program and return in the next to a fuller discussion of the problems inherent in the data analysis.

Setting up. The first task is to define a lattice of N sites, i, each of which is occupied by a spin, s_i. This needs to be done in such

Table 7.1. Flow diagram showing the steps in a Monte Carlo calculation of thermodynamic averages for a simple spin model. The system
takes n_0 steps to reach equilibrium and the total number of steps is
n_{max}

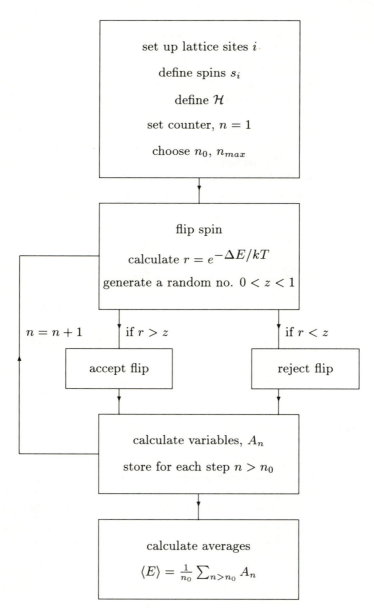

a way that a record is kept of the neighbours of each spin as its energy will be needed later. The parameters in the problem, such as the temperature and exchange interactions, should also be defined here.

Because N is necessarily finite thought must be given as to what to do with the spins on the boundaries of the system. These can either be left with fewer bonds than usual (free boundary conditions) or assumed to interact with the corresponding spin on the opposite face of the lattice (periodic boundary conditions). The latter option often gives the best results, but care must be taken that the system is not subject to false constraints. For example, simulations on a simple antiferromagnet with periodic boundary conditions can be expected to give inaccurate or spurious results if the length of the lattice is an odd number of spins.

Another consideration is the choice of initial values for the spins. Usually any choice will eventually lead to thermal equilibrium but it is helpful if this happens sooner rather than later. For a simple ferromagnet a ferromagnetically ordered state is likely to provide the most efficient initial configuration at low temperatures; at higher temperatures a random state provides the best starting point. We return to the problems of convergence to equilibrium and finite system size in Sections 7.3.1 and 7.3.3 respectively.

Generating the Markov chain. This is the heart of the program. It is summarized in the centre portion of Table 7.1. The steps are listed below

1. Select a spin, either randomly or sequentially. Calculate $r = e^{-\Delta E/kT}$ where $\Delta E = E_f - E_i$ is the change in energy associated with a possible spin flip (to a randomly chosen final state if the spin has more than two states).

2. Compare r to a random number $0 < z < 1$.

3. Flip the spin[2] if $r > z$.

4. Use the final configuration (whether the test spin was flipped or not) to generate the value of any thermodynamic quantity to be averaged. Store this value.

---

[2]It is not hard to convince oneself that this procedure reproduces the transition probability given by eqn (7.3): for $\Delta E < 0$, $r > 1$ and hence the spin is always flipped; for $\Delta E < 0$, the probability that $z < r$ is r and hence the spin is flipped with probability $r = e^{-\Delta E/kT}$.

It is important to be aware that any bias in the random number generator will introduce systematic errors into the results. The evidence is that the random number generators built into modern computers have sufficiently good statistics that the errors are insignificant compared to statistical errors. The question of how many configurations are needed to give satisfactory averages is discussed in Section 7.3.2.

Calculating the averages. Average the thermodynamic variables generated at each step of the Markov chain. Care must be taken not to include the initial states where the starting configuration still has an influence. The magnetization and energy are the easiest quantities to calculate as they are just sums over spins or products of spins.

7.3   Considerations in the data analysis

7.3.1   Influence of the starting configuration

During the first iterations of the Monte Carlo procedure the system is not in equilibrium, and hence these configurations cannot be included in the final averages. It can be hard to decide how many steps to exclude. One possibility is to perform several Monte Carlo runs with the same parameters but using different starting configurations. If the results agree to within statistical error it can be concluded that the influence of the starting configuration has been eliminated. A circumstance that can nullify this procedure, which has caused confusion in the past, is that a system can become stuck in a metastable state and feign true thermal equilibrium.

If the simulation is performed near the critical temperature, the additional problem of critical slowing down is encountered. Because of the increasing range of the correlations as criticality is approached, the time for relaxation to equilibrium τ diverges

$$\tau \sim \xi^z \tag{7.8}$$

with $z \sim 2$ for most models. In a finite system the divergence is suppressed; the smaller the system the quicker equilibrium can be achieved for a given temperature. However, at the same time finite-size corrections become more severe and a balance between these and equilibration times must be struck in the design of a Monte Carlo simulation.

An example of raw data from a Monte Carlo simulation showing the approach to equilibrium and the importance of excluding the initial

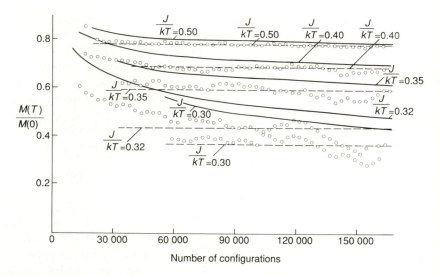

Fig. 7.1. Magnetization of an Ising ferromagnet on a cubic lattice of size $10 \times 10 \times 10$ with periodic boundary conditions plotted as a function of the number of Monte Carlo configurations for different temperatures. Open circles denote averages taken over the three preceding Monte Carlo steps per spin. Full curves give a running average if no initial configurations are excluded. The dashed lines are the final estimates of the magnetization where initial configurations have been excluded. After Binder, K. and Rauch, Z. (1969). *Zeitschrift für Physik*, **219**, 201.

configurations is shown in Fig. 7.1. Note that equilibrium is attained after a few Monte Carlo steps per spin for temperatures sufficiently far from the critical point but a slower relaxation and larger fluctuations are observed closer to T_c ($J/kT_c = 0.22$).

7.3.2   Statistical errors

To obtain reliable results for the equilibrium value of an observable, $\langle A \rangle$, the average must be taken over a time much longer than that over which the Monte Carlo states are correlated. This becomes more difficult near the critical point or if there are metastable states in the system. It can be shown that the deviation of $\hat{A}_n$ from $\langle A \rangle$ is normally distributed in the limit $n \to \infty$. Thus standard data analysis can be applied to determine the statistical error.

Dividing the equilibrium configurations into independent blocks and calculating $\hat{A}$ for each block gives a set of essentially independent estimates of $\langle A \rangle$, the variance of which gives a value for the sampling error. The problem is to know when a block of states is long enough that different blocks can indeed be considered mutually independent. A test for this is to perform the analysis using several different block sizes. The blocks are long enough when the variance becomes independent of the block size.

An alternative is to average over several different runs. The disadvantage of this procedure is that the system must be equilibrated afresh for each set of data.

7.3.3   Finite-size corrections

Because it is impossible to simulate an infinite system—for a three-dimensional Ising model $N = (128)^3$ is a realistic size on modern supercomputers— finite-size effects must be taken into account. Away from the critical point, where the correlation length is small compared to the system size, this is usually not a major problem and the parameters of the simulation can be chosen so that the errors due to the limitations in the number of spins are small compared to statistical errors.

The problem becomes much more acute as a continuous phase transition is approached, because on a finite lattice the correlation length is prevented from becoming infinite. As a result any singularities associated with the phase transition are shifted and rounded. The best compromise is to obtain high quality data for lattices of different linear dimension L, and extrapolate to $L = \infty$.

7.4   Examples

7.4.1   The three-dimensional Ising model

A lot of effort has been put into Monte Carlo simulations for the three-dimensional Ising model. This is partly because of its suitability for fast algorithms and partly because of the intrinsic interest in obtaining precise values for the critical parameters. Special purpose machines, in which the time-consuming spin updating is carried out by a specially constructed processor, have been built in several places. These are orders of magnitude cheaper than supercomputers but can only carry out the specific task for which they were designed. They can achieve an accuracy comparable to that obtained by careful programming on the most powerful conventional computers.

As an example of the run times that can be achieved we give some figures from a machine at Santa Barbara, USA[3]. This can update 10^7 spins each second. Lattice sizes of up to $N = 64 \times 64 \times 64$ were used and data for 10^7–10^8 Monte Carlo steps per spin collected for each size. At the critical temperature on the largest lattices the time to come to equilibrium was of the order of 7000 Monte Carlo steps per spin.

Using these data the result for the critical temperature was $K_c = J/kT_c = 0.221650(5)$ where the figure in brackets is the estimate of the error in the last digit. This agrees with, and is comparable in accuracy to, the best estimate from series expansions, $K_c = 0.221655(5)$[4]. The value obtained for the exponent ratio $\gamma/\nu = 1.98(2)$ is considerably less precise because of the problems of finite-size effects. Better results can be obtained using the Monte Carlo renormalization group, a technique which combines the strengths of the renormalization group and Monte Carlo simulations. This will be described in Chapter 9.

7.4.2 More complicated systems

Although Monte Carlo is particularly well suited to simulations of the Ising and other discrete spin models it was originally introduced in relation to fluids and has proved useful both here and in many other contexts. The most fundamental difference between simulations on different systems is in the choice of test configuration.

For example, for fluids, one possibility is to choose a molecule at random and allow it to move through a distance chosen at random between 0 and Δ in a random direction. The most accurate results are obtained if Δ is chosen so that approximately half the trials are accepted. Many different models have been considered in the literature, ranging from a gas of hard sphere molecules to attempts to incorporate realistic interatomic potentials. Common aims are to calculate the equation of state or the pair correlation function.

With today's computational power it is feasible to obtain realistic results for even more complicated systems. One example is solutions of polymers, long chain molecules, where Monte Carlo has been particularly useful in looking at properties which depend on the polymer topology rather than the details of the chemistry. Here the so-called reptation technique is one of the most efficient ways of generating suitable sequences of states. Starting from an arbitrary configuration the end of one of the chains is removed at random and added to the other

---

[3]Barber, M. N., Pearson, R. B., Toussaint, D., and Richardson, J. L. (1985). *Physical Review*, **B32**, 1720.

[4]Adler, J. (1983). *Journal of Physics A: Mathematical and General*, **16**, 3585.

end of the chain. As long as the self-avoidance of the chains is preserved the move is accepted according to the usual Metropolis criterion. Results have been obtained for such diverse problems as the changes in chain morphology as the temperature or solvent composition are varied, for the properties of chain molecules at surfaces, and for the dynamics of tangled polymers.

7.5 Problem

7.1 Write a Monte Carlo program to determine the temperature dependence of the energy and magnetization of a two-dimensional Ising model on a square lattice. Choose a lattice size appropriate to the power of the computer you are using. Useful illustrative results can be obtained using lattices of size as small as 6 × 6. Discuss

(i) the initial conditions used

(ii) the boundary conditions

(iii) the number of steps required to achieve thermodynamic equilibrium

(iv) error bars for the results at each temperature

(v) the effect of the finite system size.

8

The renormalization group

The approaches described so far in this book have given a broad phenomenological understanding of critical phenomena. However, although a substantial framework of results and connections has been built up, we have, as yet, no explanations for the following:

1. Continuous phase transitions fall into universality classes characterized by a given value of the critical exponents.

2. For a given universality class there is an upper critical dimension above which exponents take on mean-field values.

3. Relations between exponents, which follow as inequalities from thermodynamics, hold as equalities.

4. Critical exponents take the same value as the transition temperature is approached from above or below.

5. Two-dimensional critical exponents often appear to be rational fractions.

What is needed is a theory, based on the physics of what is happening at the critical point. We argued in Chapter 1 that the special feature of criticality is that the correlation length is infinite and that the critical system is invariant on all length scales. The aim is to write down a (hopefully short, elegant, and comprehensible) mathematical theory which embodies this physics and explains all the observations listed above. A useful theory will also allow the calculation of critical exponents and transition temperatures, if not exactly, then within an accurate and well-controlled approximation scheme.

This was the situation in 1971 when Professor K.G.Wilson of Cornell University, USA, published the first paper describing the renormalization group. His work revolutionized the field of critical phenomena and was recognized by the award of the 1982 Nobel prize for physics.

The renormalization group works by changing the length scale of a system by removing degrees of freedom. Only at criticality will the properties of the system remain unaltered by the change in scale and hence the critical behaviour is described by the so-called fixed points of the transformation. An example of this was given in Section 1.2.1. Our aim in this chapter is to explore the behaviour of a renormalization group in the vicinity of its fixed points. This will lead to the important result that the thermodynamic functions can be written in a scaling form. The equality of the critical exponents above and below the critical temperature and the relations between them then follow immediately. We shall also describe how the division into universality classes can be explained by considering how the system behaves under repeated iterations of the renormalization group.

In the next chapter we consider the renormalization group as a calculational tool. Various methods of implementing the scale change are discussed and fixed points and exponents calculated. Only in very simple cases do these follow exactly from the transformation procedure. Usually approximations must be introduced which often have the drawback that, although they can work well, they are uncontrolled: there is no small parameter which allows an estimate of the error.

8.1   Definition of a renormalization group transformation

The first step is to define a renormalization group transformation and to introduce the concept of the fixed points of the mapping which will describe systems at criticality. The initial model, described by a reduced Hamiltonian, $\bar{\mathcal{H}} \equiv \mathcal{H}/kT$, is renormalized to a new system, described by a new reduced Hamiltonian $\bar{\mathcal{H}}'$:

$$\bar{\mathcal{H}}' = \mathbf{R}\bar{\mathcal{H}}. \tag{8.1}$$

The renormalization group operator $\mathbf{R}$ decreases the number of degrees of freedom from N to N'. This can be done either in real space, by removing or grouping spins, or in reciprocal space, by integrating out large wavevectors. The scale factor of the transformation, b, is defined by

$$b^d = N/N'. \tag{8.2}$$

An example of a renormalization transformation is shown in Fig. 8.1.

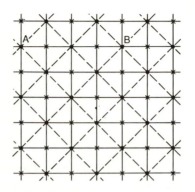

Fig. 8.1. Renormalization of the square lattice by a scale factor $b = \sqrt{2}$. The original and renormalized lattices are drawn using full and dashed lines respectively. The points A and B are 4 units apart on the original lattice but $4/b$ units apart on the renormalized lattice because lengths are measured in units of the lattice spacing.

A partial trace in the partition function is performed over the spins represented by a cross leaving a Hamiltonian which depends on only half the original spins. The new lattice retains its original topology but with a change in scale of $b = \sqrt{2}$. This procedure cannot be carried out exactly: approximation methods are described in the next chapter.

The essential condition to be satisfied by any renormalization group transformation is that the partition function must not change:

$$\mathcal{Z}_{N'}(\bar{\mathcal{H}}') = \mathcal{Z}_N(\bar{\mathcal{H}}). \tag{8.3}$$

Therefore the total free energy remains the same and, because the free energy is extensive, the reduced free energy per spin, $\bar{f} = f/kT$, transforms as

$$\bar{f}(\bar{\mathcal{H}}') = b^d \bar{f}(\bar{\mathcal{H}}). \tag{8.4}$$

This is an important equation which will lead us to a scaling form for the free energy.

Lengths, which are now measured in terms of the new lattice spacing, are reduced by a factor b. For example, the points A and B in Fig. 8.1 are 4 lattice spacings apart on the original lattice but $4/\sqrt{2}$ spacings apart on the renormalized lattice,

$$\vec{r} \Rightarrow \vec{r}\,' = b^{-1}\vec{r}. \tag{8.5}$$

Similarly momenta, which have the dimensions of inverse length, are renormalized according to

$$\vec{q} \Rightarrow \vec{q}\,' = b\vec{q}. \qquad (8.6)$$

To obtain the correct behaviour of the spin correlation function it is also necessary to rescale spin vectors as

$$\vec{s}_{\vec{r}} \Rightarrow \vec{s}\,'_{\vec{r}'} = c^{-1}\vec{s}_{\vec{r}}. \qquad (8.7)$$

Therefore the pair correlation function, which depends on a product of two spins, transforms as

$$c^2\Gamma(b^{-1}\vec{r}, \bar{\mathcal{H}}') = \Gamma(\vec{r}, \bar{\mathcal{H}}). \qquad (8.8)$$

We shall later identify c in terms of b and eqn (8.8) will lead to a scaling form for the pair correlation function and the scaling laws which involve η and ν.

The aim will be to find the fixed points of the renormalization transformation (eqn 8.1),

$$\bar{\mathcal{H}}' = \bar{\mathcal{H}} \equiv \bar{\mathcal{H}}^* \qquad (8.9)$$

where the system is invariant under a scale change. We have argued that scale invariance is a signature of criticality. To reinforce the connection consider the behaviour of the correlation length. At a fixed point the correlation length, measured in units of the lattice spacing, must map on to itself

$$\xi' = \xi \equiv \xi^*. \qquad (8.10)$$

But, because of the scale change (8.5), the correlation length, in parallel with all the lengths in the problem, is reduced by a factor b

$$\xi' = b^{-1}\xi. \qquad (8.11)$$

Equations (8.10) and (8.11) are consistent only if the correlation length is infinite (or trivially zero), as expected at criticality.

8.2 Flows in parameter space

The next step is to look more closely at the behaviour of the Hamiltonian in the vicinity of the fixed points of a renormalization group transformation. To do this we introduce the concept of parameter space and show that a renormalization transformation corresponds to flows in this space. A general reduced Hamiltonian can be written

$$\bar{\mathcal{H}} = \sum_{\alpha} \vec{\mu}.\vec{f} \qquad (8.12)$$

where the components of $\vec{f}$ are products of operators of suitable symmetry and the corresponding components of $\vec{\mu}$ are the conjugate fields.

For example, for the spin-1/2 Ising model, the components of $\vec{f}$ will include a single-spin term coupling to the magnetic field, pair interactions between first, second, and further-neighbours, three-spin interactions, and so forth. Even if the initial Hamiltonian contains only a few terms, iteration of the renormalization group transformation will generate an infinite number of multi-spin couplings between all possible neighbours.

The fields $\vec{\mu}$ can be considered as a vector marking the position of the system in an infinite-dimensional parameter space. Under successive applications of a renormalization group transformation the Hamiltonian changes and the system moves through parameter space:

$$\vec{\mu}' = \mathbf{R}\vec{\mu}. \tag{8.13}$$

At a fixed point

$$\vec{\mu}' = \vec{\mu} \equiv \vec{\mu}^*. \tag{8.14}$$

To expand around the fixed point we write

$$\vec{\mu} = \vec{\mu}^* + \delta\vec{\mu}, \tag{8.15}$$

$$\vec{\mu}' = \vec{\mu}^* + \delta\vec{\mu}'. \tag{8.16}$$

The small deviations from the fixed point in the original and renormalized systems can be related by performing a Taylor expansion on the renormalization group transformation (8.13)

$$\delta\vec{\mu}' = \mathbf{A}(\vec{\mu}^*)\delta\vec{\mu} \tag{8.17}$$

where the constant matrix $\mathbf{A}$ is evaluated at the fixed point $\vec{\mu}^*$. The eigenvalues λ_i and the corresponding eigenvectors $\vec{v}_i$ of $\mathbf{A}$ are important in determining the critical properties of the Hamiltonian.

The eigenvalues λ_i are functions of b, the scale change of the system under renormalization. If two successive transformations with scale factors b_1, b_2 are performed the total change in scale is $b_1 b_2$. It follows that

$$\lambda_i(b_1)\lambda_i(b_2) = \lambda_i(b_1 b_2). \tag{8.18}$$

This constrains the eigenvalues to be of the form

$$\lambda_i(b) = b^{y_i} \tag{8.19}$$

where y_i is independent of the scale factor, b. The y_i are critical exponents and we shall see in Section 8.5 how they are closely related to $\alpha, \beta, \gamma, \ldots$. But we first describe how they determine the flows in parameter space near the fixed points and how this leads to the concept of a critical surface and an explanation of the universality of critical exponents.

For a Hamiltonian near a fixed point, $\vec{\mu}^*$, the deviation from the fixed point may be expanded in terms of the eigenvectors of $\mathbf{A}$, $\vec{v}_i$,

$$\vec{\mu} = \vec{\mu}^* + \sum_i g_i \vec{v}_i. \tag{8.20}$$

The coefficients g_i are termed the linear scaling fields. Under renormalization

$$\vec{\mu}' = \vec{\mu}^* + \sum_i b^{y_i} g_i \vec{v}_i. \tag{8.21}$$

or, more succinctly,

$$g_i' = b^{y_i} g_i. \tag{8.22}$$

It follows from eqn (8.22) that the flow of a Hamiltonian, originally close to the fixed point, within parameter space depends on the set of scaling fields g_i describing the original position of the Hamiltonian and on the form of the corresponding y_i. For a positive y_i the scaling field g_i increases under repeated iterations of the renormalization transformation and drives the system away from the fixed point. (Exactly how we cannot tell, as the assumption of linearity breaks down.) This is called a relevant variable. Unless all the relevant scaling fields are initially equal to zero the system must eventually flow away from the fixed point.

If the exponent y_i is negative the corresponding scaling field decreases under repeated iterations of the renormalization group, thus moving the system closer to the fixed point. These scaling fields are termed irrelevant variables. Whatever their initial value the fixed point will still be attained.

If any y_i is zero the corresponding scaling field is termed a marginal variable and higher order terms become important. This case, which allows a continuous variation of exponents with an interaction parameter in the Hamiltonian[1], will not concern us further here.

Thus the stability of the fixed point depends on the numbers of relevant and irrelevant eigenvalues associated with it. To illustrate this it is helpful to plot flow diagrams showing the trajectories in parameter space along which a system flows under repeated iteration of a renormalization group transformation. Fig. 8.2 shows a fixed point

---

[1] Nienhuis, B. (1987). *Coulomb gas formulation of two-dimensional phase transitions*. In *Phase transitions and critical phenomena*, Vol 11 (eds C. Domb and J. L. Lebowitz), p.1. (Academic Press, London).

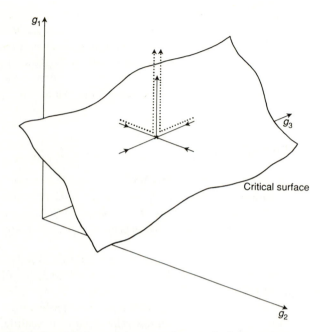

Fig. 8.2. A fixed point with one relevant (g_1) and two irrelevant (g_2 and g_3) scaling fields. The two-dimensional critical surface whence all trajectories flow to the fixed point is defined locally by $g_1 = 0$. The dotted trajectories start close to, but not on, the critical surface. They initially flow towards the fixed point but are eventually driven away from it by the increase of the relevant scaling field.

with one relevant and two irrelevant scaling fields. Any system which corresponds to an initial position in parameter space, $\vec{\mu}$, which has a component along the relevant eigenvector must eventually be driven away from the fixed point by the increase of the relevant scaling field. The dotted trajectories are examples of this behaviour. Physically this corresponds to the system moving away from criticality as it is observed on increasingly long length scales. Examples are shown in Figs 1.8 and 1.10.

The points which flow into the fixed point define a surface on which the relevant scaling field is zero. This surface is called the critical surface. It corresponds to all points representing systems with an infinite correlation length. This follows immediately from noting that the flow is to the fixed point, which itself has an infinite correlation length, and that the correlation length can only decrease under the application of a renormalization group transformation.

To define the critical surface it is usually necessary to control only a small number of relevant parameters. For example, for most magnets with short-ranged interactions criticality is achieved if the magnetic field and temperature are tuned to the correct values. Thus the critical surface has two dimensions fewer than the space of all coupling constants considered.

8.3   Universality

We have throughout this book stressed the importance of universality, the independence of the values of the critical exponents of the detailed interactions in a system. How does this follow from the renormalization group picture of flows through parameter space under successive changes in length scale?

Consider a critical surface which contains only one non-trivial fixed point. All systems which are close to criticality will initially lie close to the critical surface in parameter space. Their position can be expanded in terms of a set of (in general non-linear) relevant and irrelevant scaling fields. Under a scale change the irrelevant scaling fields will decrease and the system will flow towards the fixed point, while the relevant fields will increase, driving it away from the critical surface. As long as the relevant fields are initially small enough the trajectory will come close to the fixed point before it starts significantly to move away from the critical surface. Examples are the dotted trajectories in Fig. 8.2. Therefore its critical behaviour will be determined by the linearized transformation at the fixed point and will be independent of the original values of the irrelevant scaling fields. Thus all systems which flow close to the fixed point, independent of where in parameter space they start to move, will exhibit the same critical exponents determined by the eigenvalues of the linear transformation matrix at that fixed point. This is the property of universality.

8.3.1   Crossover

If there is more than one fixed point embedded in a critical surface crossover effects may occur. For example, in a magnetic system with weak spin anisotropy as the temperature approaches T_c, the system exhibits Heisenberg critical behaviour, but very close to T_c the critical exponents change to those corresponding to an Ising system. In parameter space this behaviour corresponds to a trajectory which passes close to the Heisenberg fixed point but is eventually driven away by a

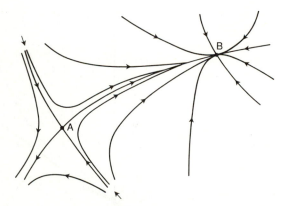

Fig. 8.3. Part of a critical surface containing two fixed points, A and B. A has one relevant and one irrelevant and B two irrelevant scaling fields. For the trajectories marked by an arrow the relevant scaling field is initially very small so the trajectories pass close to A before being driven to B by the increase of the field. Hence, as the critical temperature is approached a crossover is seen between the critical behaviour described by the fixed point A and that described by B.

relevant interaction to the Ising fixed point, where the interaction is no longer relevant. In this example the extra relevant variable at the Heisenberg fixed point is the anisotropic part of the exchange interaction. An example of crossover is shown in Fig. 8.3.

8.4   An example

We leave discussing the difficulties and approximations inherent in deriving specific renormalization group recursion equations to the next chapter. However, it is useful to present here a concrete example of the ideas discussed in the preceding sections. Therefore we write down, without proof, a set of realistic recursion equations for a renormalization group with scale parameter $b = 2$ which describe flows within a two-dimensional parameter space:

$$y' = \frac{4y^2}{(1 + y^2)^2}, \qquad z' = \frac{z^4}{(1 + y^2)^2}. \qquad (8.23)$$

These equations are derived in problem 9.4.

Putting $(y', z') = (y, z)$ and solving the resulting equations gives fixed points at $(y^*, z^*) = (0, 0), (0, 1), (0, \infty), (0.296, 0), (0.296, 1.057),$

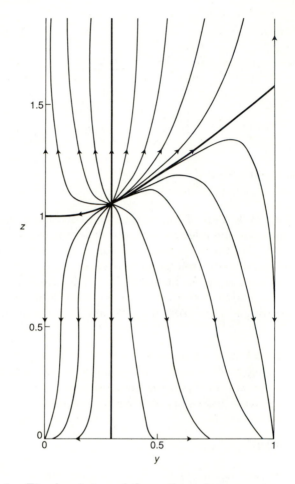

Fig. 8.4. Fixed points and flows through parameter space for the recursion eqns (8.23).

$(0.296, \infty)$, $(1, 0)$, $(1, 1.587)$, and $(1, \infty)$ where we have considered only $0 \le y \le 1$ and $0 \le z \le \infty$. These are plotted in Fig. 8.4 together with the flows through parameter space which follow from iterating the recursion equations.

Linearizing eqn (8.23) about a given fixed point gives the matrix **A** of eqn (8.17)

$$\begin{pmatrix} \delta y' \\ \delta z' \end{pmatrix} = \begin{pmatrix} 8y^*(1 - y^{*2})/(1 + y^{*2})^3 & 0 \\ -z^{*4}/(1 + y^{*2})^3 & 4z^{*3}/(1 + y^{*2})^2 \end{pmatrix} \begin{pmatrix} \delta y \\ \delta z \end{pmatrix}$$

$$(8.24)$$

with eigenvalues

$$\lambda_1 = 8y^*(1 - y^{*2})/(1 + y^{*2})^3, \qquad \lambda_2 = 4z^{*3}/(1 + y^{*2})^2. \quad (8.25)$$

For the fixed point at $(0.296, 1.057)$, $\lambda_1 = 1.679$ and $\lambda_2 = 4$. These are both greater than unity, so the corresponding values of y_i, defined by eqn (8.19), are both positive. Therefore this fixed point has two relevant scaling fields and in a two-dimensional parameter space is unstable to any perturbation away from the fixed-point value. A glance at Fig. 8.4 will verify that this is indeed the case. For the fixed point at $(0.296, 0)$ one relevant and one irrelevant scaling field are expected. The eigenvalues here are $\lambda_1 = 1.679$ and $\lambda_2 = 0$. Hence y_1 is positive and relevant, whereas y_2 is negative and irrelevant and the critical surface which forms the basin of attraction for this fixed point is one-dimensional.

8.5   Scaling and critical exponents

All the formalism is now in place to enable us to see how the singular part[2] of the free energy and the correlation function behave under a renormalization transformation. We shall show that they belong to a special class of functions that can be written in scaling form. As a consequence of this it will be possible to write down the critical exponents, $\alpha, \beta, \gamma, \ldots$, in terms of the exponents, y_i, which are directly related through eqn (8.19) to the eigenvalues of the linearized transformation matrix at the fixed point. Relations between the critical exponents will then follow immediately.

The original and renormalized Hamiltonians, $\bar{\mathcal{H}}$ and $\bar{\mathcal{H}}'$, correspond to positions in parameter space, $\vec{\mu}$ and $\vec{\mu}'$. So eqn (8.4), which describes the renormalization of the singular part of the reduced free energy per spin, may be written

$$\bar{f}_s(\vec{\mu}) = b^{-d}\bar{f}_s(\vec{\mu}'). \quad (8.26)$$

Near a fixed point $\vec{\mu}$ and $\vec{\mu}'$ can be written in terms of the linear scaling fields, g_i and g'_i, which are related in the original and renormalized models by eqn (8.22). Hence eqn (8.26) becomes

---

[2]At present we ignore the renormalization of the constant term in the Hamiltonian. See Section 9.1 for an explicit example of how this behaves.

$$\bar{f}_s(g_1, g_2, g_3 \ldots) \sim b^{-d} \bar{f}_s(b^{y_1} g_1, b^{y_2} g_2, b^{y_3} g_3 \ldots). \tag{8.27}$$

The singular part of the free energy near a fixed point has been written in so-called scaling form. Equation (8.27) is true for arbitrary b since by repeated iterations of the transformation we may make b as large as we like. This is an example of a generalized homogeneous function. A description of these functions and some examples are given in problem 8.1.

To relate the y_i to the critical exponents defined in Table 2.3 we need to identify the linear scaling fields g_i. We have argued that in most magnets with short-range interactions there are two relevant variables, as two parameters have to be adjusted to attain criticality. These are the reduced temperature t defined by eqn (2.18) and the reduced magnetic field, $h = H/kT$. Therefore, choosing $g_1 = t$ and $g_2 = h$, and assuming that all other scaling fields are irrelevant, eqn (8.27) becomes

$$\bar{f}_s(t, h, g_3 \ldots) \sim b^{-d} \bar{f}_s(b^{y_1} t, b^{y_2} h, b^{y_3} g_3 \ldots) \tag{8.28}$$

as $t, h, g_3 \to 0$.

Recall from Table 2.3 the definition of the exponent α which describes how the zero-field specific heat varies with temperature on approaching the critical point

$$C \sim \left(\frac{\partial^2 \bar{f}_s}{\partial t^2} \right)_{h=0} \equiv \bar{f}_{tt}(h = 0) \sim | t |^{-\alpha} . \tag{8.29}$$

Differentiating eqn (8.28) twice with respect to temperature and putting h and the irrelevant variables equal to zero gives

$$\bar{f}_{tt}(t, 0) \sim b^{-d + 2y_1} \bar{f}_{tt}(b^{y_1} t, 0). \tag{8.30}$$

The aim is to extract the temperature dependence of the right-hand side of eqn (8.30). This can be done because the rescaling factor b is arbitrary. Choosing $b^{y_1} |t| = 1$ transfers all temperature dependence to the prefactor and leaves it multiplied by a function of constant argument

$$\bar{f}_{tt}(t, 0) \sim | t |^{(d - 2y_1)/y_1} \bar{f}_{tt}(\pm 1, 0) \tag{8.31}$$

and we may identify

$$\alpha = 2 - d/y_1. \tag{8.32}$$

Similarly (see problem 8.2)

$$\beta = (d - y_2)/y_1, \tag{8.33}$$

$$\gamma = (2y_2 - d)/y_1, \tag{8.34}$$
$$\delta = y_2/(d - y_2). \tag{8.35}$$

We have succeeded in writing four critical exponents in terms of two variables y_1 and y_2. Hence there will be two relations between the exponents. It is easily checked that

$$\alpha + 2\beta + \gamma = 2, \tag{8.36}$$

$$\gamma = \beta(\delta - 1). \tag{8.37}$$

These should be compared to the exponent inequalities (2.27) and (2.29). It also follows immediately that the exponents describing the behaviour of a given thermodynamic function are the same above and below the critical temperature; for example, the leading behaviour of $\bar{f}_{tt}$ in eqn (8.31) does not depend on the sign of t.

Expressions for the exponents ν and η follow from a similar scaling form for the pair correlation function (eqn 8.8). Near a fixed point this becomes

$$\Gamma(\vec{r}, t, h, g_3 \ldots) \sim c^2(b)\Gamma(b^{-1}\vec{r}, b^{y_1}t, b^{y_2}h, b^{y_3}g_3 \ldots). \tag{8.38}$$

Putting h and the irrelevant variables to zero and choosing $b^{y_1}|t| = 1$

$$\Gamma(\vec{r}, t) \sim c^2(|t|^{-1/y_1})\Gamma(|t|^{1/y_1}\vec{r}, \pm 1). \tag{8.39}$$

Lengths, including the correlation length, scale as $b \sim |t|^{-1/y_1}$ and hence

$$\nu = 1/y_1. \tag{8.40}$$

Equations (8.40) and (8.32) can be combined to give the hyperscaling relation

$$2 - \alpha = d\nu. \tag{8.41}$$

To identify c^2 recall that for large r at the critical point

$$\Gamma(\vec{r}) \sim r^{-(d-2+\eta)}. \tag{8.42}$$

Putting t, h, and the irrelevant variables equal to zero in eqn (8.38) it is apparent that eqns (8.38) and (8.42) are consistent only if

$$c^2(b) = b^{-(d-2+\eta)/2}. \tag{8.43}$$

To relate η to the other exponents we shall use eqn (2.17):

$$\chi_T \sim N \int \Gamma(r) r^{d-1} dr. \tag{8.44}$$

Using eqn (8.42) and making the substitution $r = x\xi$

$$\chi_T \sim \xi^{2-\eta}. \tag{8.45}$$

Writing both sides in terms of t using the definitions of γ and ν

$$t^{-\gamma} \sim t^{-(2-\eta)\nu}. \tag{8.46}$$

Therefore

$$\gamma = (2 - \eta)\nu. \tag{8.47}$$

This equation should be compared to eqn (2.29).

8.6 Scaled variables

The thermodynamic functions may be plotted in a way that explicitly displays their scaling form by suppressing one of the variables, say the temperature. Differentiating eqn (8.28) with respect to the magnetic field gives a scaling form for the magnetization M:

$$M(t, h, g_3 \ldots) \sim b^{-d+y_2} M(b^{y_1} t, b^{y_2} h, b^{y_3} g_3 \ldots). \tag{8.48}$$

Choosing $b^{y_1} |t| = 1$ and putting the irrelevant variables equal to zero,

$$M(t, h) \sim |t|^{(d-y_2)/y_1} M(\pm 1, h \, |t|^{-y_2/y_1}). \tag{8.49}$$

Equations (8.33) and (8.35) can now be used to replace y_1 and y_2 by β and δ:

$$M(t, h) \sim |t|^{\beta} M(\pm 1, h \, |t|^{-\beta\delta}). \tag{8.50}$$

Defining the scaled magnetization and field:

$$\tilde{m} = \frac{M(t, h)}{|t|^{\beta}}; \qquad \tilde{h} = h \, |t|^{-\beta\delta}, \tag{8.51}$$

eqn (8.50) becomes

$$\tilde{m} \sim M(\pm 1, \tilde{h}). \tag{8.52}$$

Hence a plot of $\tilde{m}$ against $\tilde{h}$ should lie on the same curve for all values of temperature greater than or less than the critical temperature

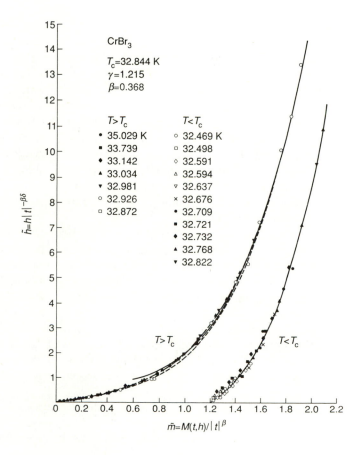

Fig. 8.5. Plot of scaled magnetic field versus scaled magnetization for the ferromagnet, CrBr$_3$. Note the collapse of the data on to two curves, one for temperatures greater and one for temperatures less than the critical temperature. After Ho, J. T. and Lister, J. D. (1969). *Physical Review Letters*, **22**, 603.

respectively, provided that they lie within the scaling regime. Fig. 8.5 gives an example for the ferromagnet $CrBr_3$[3].

8.7 Conformal invariance

Many of the questions posed at the beginning of this chapter now have an answer. However, nowhere in the renormalization group formalism is there an explanation of why two-dimensional critical exponents are often rational fractions. An understanding of this came somewhat later, in 1984, when the import of the fact that systems at criticality are usually not only scale invariant but also conformally invariant was recognized. In a conformal transformation the length rescaling factor, $b(\vec{r})$, can depend continuously on position, as shown in Fig. 8.6. The transformation must look locally like a combination of a dilation, a rotation, and a translation: no shear components are allowed. Locally the rescaled lattice retains its shape.

Conformal invariance has proved particularly powerful in two dimensions because the conformal group is much larger than in higher dimensions. It has been used to show that the allowed values of the critical exponents are restricted to certain simple rational numbers which have been identified, by trial and error, as pertaining to particular models[4].

Conformal invariance can also be exploited to relate the properties of spin models on strips of finite width L and infinite length with periodic boundary conditions to those of the corresponding two-dimensional system. This is a consequence of the conformal mapping

$$w = \frac{L}{2\pi} \ln z \tag{8.53}$$

which maps the entire z-plane on to the surface of a cylinder. One

---

[3]We have shown that the renormalization group leads to scaling forms for the thermodynamic functions and hence that experimental data collapse on to two curves when plotted in appropriate scaled variables. However, as can be seen from the date on which the data presented in Fig. 8.5 were taken, the experiments and the ideas of scaling did in fact pre-date the renormalization group. An early attempt to explain scaling, which foreshadowed many of the ideas of the renormalization group, was due to Kadanoff, L. P. (1966). *Physics*, **2**, 263.

[4]This takes sophisticated mathematics. See, for example, Cardy, J. L. (1987). *Conformal invariance.* In *Phase transitions and critical phenomena*, Vol 11 (eds C. Domb and J. L. Lebowitz), p.55. (Academic Press, London).

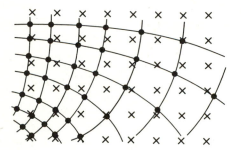

Fig. 8.6. A conformal transformation of the square lattice. After Cardy, J. L. (1987). *Conformal invariance.* In *Phase transitions and critical phenomena*, Vol 11 (eds C. Domb and J. L. Lebowitz), p.55. (Academic Press, London).

particularly simple result is that at the critical temperature, as $L \to \infty$, the correlation length is related to the critical exponent η by

$$\xi = \frac{L}{\pi\eta}. \tag{8.54}$$

For example, for the two-dimensional Ising model, it is known exactly that

$$\xi_L = \frac{4}{\pi}L(1 - \pi^2/24L^2 + \ldots) \tag{8.55}$$

giving $\eta = 1/4$ as expected.

The relation between strips and two-dimensional systems is particularly useful in exploring the critical properties of the latter[5] because discrete spin models on strips can be described by a finite transfer matrix ($2^L \times 2^L$ for the Ising model on a strip of width L). There are many routines available to diagonalize large matrices and, as it is often only necessary to calculate the few largest eigenvalues, matrices of sizes up to $10^5 \times 10^5$ can be handled without too much trouble. This corresponds to $L \sim 16$ for the two-dimensional Ising model.

8.8 Problems

8.1 A function $f(x, y)$ is defined to be a generalized homogeneous function if

---

[5]Several examples are given in Cardy, J. L. (ed.) (1988). *Finite-size scaling.* (North-Holland, Amsterdam).

Table 8.1. Magnetization, in μ_B per atom, of nickel. The data are taken from Weiss, P. and Forrer, R. (1926). *Annales de Chimie et de Physique*, **5**, 153

Magnetic field ($\times 10^4$ A m^{-1})	Temperature(K)			
	621.9	623.8	625.7	631.3
3.42	0.119	0.110	0.085	—
7.08	0.145	—	—	—
10.78	0.150	0.133	0.112	0.044
25.70	0.164	0.150	0.134	0.079
47.87	0.178	0.166	0.152	0.109
80.13	0.191	0.181	0.170	0.134
113.1	0.202	0.193	0.183	0.151
141.4	0.210	0.201	0.192	0.163

$$f(\lambda^a x, \lambda^b y) = \lambda f(x, y)$$

for all values of the parameter, λ.
(i) Prove that

$$(a) \qquad f(x, y) = x^3 + y^2$$
$$(b) \qquad f(x, y) = x^3 y^2 + x^2 y^3$$

are generalized homogeneous functions and find the corresponding values of a and b.
(ii) Prove that

$$f(x, y) = x^{1/a} \tilde{f}(x^{-b/a} y) = y^{1/b} \tilde{\tilde{f}}(y^{-a/b} x).$$

(iii) Find the functions $\tilde{f}$ and $\tilde{\tilde{f}}$ for the examples given in (i).

8.2 Starting from the scaling form for the singular part of the free energy

$$\bar{f}_s(t, h) \sim b^{-d} \bar{f}_s(b^{y_1} t, b^{y_2} h)$$

where $t = (T - T_c)/T_c$ and $h = H/kT$ show that

$$\alpha = 2 - d/y_1, \qquad \beta = (d - y_2)/y_1,$$
$$\gamma = (2y_2 - d)/y_1, \qquad \delta = y_2/(d - y_2)$$

and hence confirm that

$$\alpha + 2\beta + \gamma = 2, \qquad \gamma = \beta(\delta - 1).$$

8.3[6] (i) Show that the scaling hypothesis

$$\bar{f}_s(t, h) \sim b^{-d} \bar{f}_s(b^{y_1} t, b^{y_2} h)$$

may be rewritten, for a ferromagnet, in Griffith's form

$$h \sim \text{sgn}\{m\} \mid m \mid^{\delta} Z(t/ \mid m \mid^{1/\beta})$$

for small m, h, and t.
(ii) Use this result to show that

$$\gamma = \beta(\delta - 1).$$

8.4 Using the variables[7] $x = e^{-4J/kT}$ and $h = H/kT$ check that, for small x and h, the free energy and magnetization of the one-dimensional Ising model given in eqns (5.32) and (5.35) can be recast in scaling form. What are the values of y_1 and y_2?

8.5 Table 8.1 lists early measurements of the magnetization of nickel. To what extent do these results support the scaling hypothesis? For nickel $T_c = 627.2K$, $\beta = 0.368$, and $\delta = 4.22$.

---

[6] After M. E. Fisher.
[7] The variable x, rather than $t = (T - T_c)/T_c$, is appropriate for the one-dimensional Ising model because $\xi \sim x^{-1/y_1}$ rather than $\xi \sim t^{-1/y_1}$. This is a consequence of the zero temperature phase transition.

9

Implementations of the renormalization group

In Chapter 8 we gave an introduction to the formalism of the renormalization group. A renormalization transformation changes the scale of a system by a factor b. Systems at criticality flow to fixed points under the transformation and their critical exponents and scaling behaviour are determined by linearizing the recursion equations about the fixed point.

The aim of this chapter is two-fold: firstly to illustrate these ideas by giving several examples of renormalization group transformations and secondly to demonstrate the ways in which they can be used to calculate critical properties. Our examples are divided into three classes. In Section 9.1, we shall consider a one-dimensional model where the recursion relations can be written down exactly. The drawback of this as a canonical example is that any critical behaviour is at zero temperature. The decimation transformation presented in one dimension is an example of a real-space transformation and the second task, in Section 9.2, is to show how these may be used approximately to treat models in higher dimensions. Thirdly, in Section 9.5, we shall mention the ϵ-expansion where the renormalization takes place in momentum space. This enables an understanding of the crossover to mean-field behaviour. However, the formalism is somewhat involved and the details are beyond the scope of this book.

9.1 The one-dimensional Ising model

An exact renormalization group transformation can be written down for the one-dimensional Ising model in a simple way. Most of the important features of the renormalization group are already apparent and it will prove a helpful example to illustrate some of the main ideas before

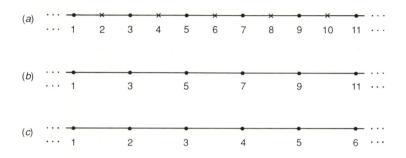

Fig. 9.1. Renormalization of the one-dimensional Ising model by a scale factor, $b = 2$: (a) the original lattice, (b) the renormalized lattice, (c) the renormalized lattice with the spins re-numbered consecutively.

becoming embroiled in the intricacies of the approximations needed in higher dimensions.

The first step is to write down the recursion equations for the one-dimensional Ising model in a magnetic field and to extract the fixed point structure. We shall then show how a scaling form for the free energy results from studying the behaviour of the recursion equations near the non-trivial fixed point. One problem of using the one-dimensional Ising model as an example is that the transition is at zero temperature: this necessitates care in the choice of scaling variables and the definitions of the critical exponents. We shall also show that by iterating the recursion equations the free energy can be calculated for all temperatures.

9.1.1   Derivation of the recursion equations

The reduced Hamiltonian for the one-dimensional Ising model is, as usual,

$$\bar{\mathcal{H}} = -K \sum_{\langle ij \rangle} s_i s_j - h \sum_i s_i - \sum_i C \tag{9.1}$$

where the constant or background term, C, is included because, even if set to zero at this stage, it will be generated by the renormalization transformation.

The aim is to change the scale of the system and we shall do this by a factor $b = 2$ by performing the trace in the partition function over the even-numbered spins. This procedure, illustrated in Fig. 9.1, is appropriately termed decimation. The partition function can be written as a product of terms each of which depends on only one of the even-numbered spins

$$\mathcal{Z} = \sum_{\{s\}} \prod_{i=\ldots 2,4,6\ldots} \exp\{Ks_i(s_{i-1}+s_{i+1})+hs_i+h(s_{i-1}+s_{i+1})/2+2C\}.$$

(9.2)

Hence doing the partial trace is easy and gives

$$\mathcal{Z}' = \sum_{\ldots s_1,s_3,s_5\ldots} \prod_{i=\ldots 2,4,6\ldots} \{\exp[K(s_{i-1}+s_{i+1})+h+h(s_{i-1}+s_{i+1})/2+2C]$$

$$+ \exp[-K(s_{i-1}+s_{i+1})-h+h(s_{i-1}+s_{i+1})/2+2C]\}. \qquad (9.3)$$

Relabelling the spins so they are numbered consecutively—this is a matter of convenience only—see Fig. 9.1

$$\mathcal{Z}' = \sum_{\{s\}} \prod_i \{\exp[(K+\frac{h}{2})(s_i+s_{i+1})+h+2C]$$

$$+ \exp[-(K-\frac{h}{2})(s_i+s_{i+1})-h+2C]\}. \qquad (9.4)$$

We demand that the renormalized partition function can be cast in the same form as that of the original system but with renormalized coupling constants

$$\mathcal{Z}' = \sum_{\{s\}} \prod_i \exp(K's_is_{i+1}+h's_i+C'). \qquad (9.5)$$

One sees immediately that 9.4 and 9.5 are consistent if

$$\exp\{K's_is_{i+1}+h'(s_i+s_{i+1})/2+C'\} =$$

$$\exp\{(K+\frac{h}{2})(s_i+s_{i+1})+h+2C\}$$

$$+ \exp\{-(K-\frac{h}{2})(s_i+s_{i+1})-h+2C\} \qquad (9.6)$$

for $s_i, s_{i+1} = \pm 1$. This leads to three equations

$$
\begin{aligned}
s_i = s_{i+1} = 1: \quad & e^{K'+h'+C'} = e^{2K+2h+2C} + e^{-2K+2C}, \\
s_i = s_{i+1} = -1: \quad & e^{K'-h'+C'} = e^{2K-2h+2C} + e^{-2K+2C}, \\
s_i = -s_{i+1} = \pm 1: \quad & e^{-K'+C'} = e^{h+2C} + e^{-h+2C}
\end{aligned}
$$

(9.7)

and hence expressions for the renormalized parameters

$$e^{2h'} = e^{2h}\cosh(2K+h)/\cosh(2K-h), \qquad (9.8)$$

$$e^{4K'} = \cosh(2K+h)\cosh(2K-h)/\cosh^2 h, \qquad (9.9)$$

$$e^{4C'} = e^{8C}\cosh(2K+h)\cosh(2K-h)\cosh^2 h. \qquad (9.10)$$

We should stress that the special feature of the one-dimensional Ising model which makes it amenable to an exact renormalization group solution is that the renormalized partition function *can* be written in the same form as the original one. More generally new further-neighbour and multi-spin interactions must be introduced at each step of a renormalization group transformation and some way needs to be found of coping with the proliferation of couplings. This is discussed further in Section 9.2.

A final point before we analyse the transformation 9.8–9.10: recursion equations for one-dimensional models can be generated more easily but less explicitly using the transfer matrix formalism. Examples are given in problems 9.1 and 9.3.

9.1.2   Fixed points

To analyse the recursion equations it is convenient to work with a new set of variables

$$\omega = e^{-4C}, \quad x = e^{-4K}, \quad y = e^{-2h}. \qquad (9.11)$$

In terms of these eqns (9.8)–(9.10) become

$$\omega' = \omega^2 x y^2 (1+y)^2/(x+y)(1+xy), \qquad (9.12)$$

$$x' = x(1+y)^2/(x+y)(1+xy), \qquad (9.13)$$

$$y' = y(x+y)/(1+xy). \qquad (9.14)$$

The recursion equations for x and y do not depend on the spin-independent term, ω. This is a general feature of the renormalization group: a change in the energy scale cannot alter the singular behaviour of the free energy. Therefore we can initially ignore the recursion equation for ω. It will be important when calculating the free energy for temperatures outside the critical region.

The fixed points and flows within the (x,y) plane are shown in Fig. 9.2. We restrict our attention to ferromagnetic coupling ($K > 0$) and positive magnetic fields. There is a line of fixed points at $x^* = 1$, $0 \le y^* \le 1$. $x = 1$ corresponds to infinite temperature and these are paramagnetic fixed points or sinks: all trajectories which start at positive values of x or non-zero temperature flow to them. This is to be expected because for all positive temperatures the one-dimensional

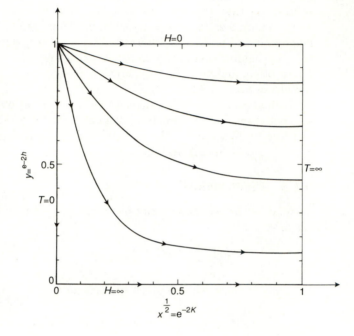

Fig. 9.2. Trajectories in parameter space for the recursion relations arising from the decimation of a one-dimensional Ising chain by a scale factor $b = 2$. There are fixed points at $x^* = 0$, $y^* = 1$ corresponding to criticality, at $x^* = y^* = 0$ corresponding to a fully aligned configuration, and at $x^* = 1$, $0 \leq y^* \leq 1$ corresponding to an infinite temperature sink. After Nelson, D. R. and Fisher, M. E. (1975). *Annals of Physics*, **91**, 226.

Ising model is above its transition temperature and must therefore look completely disordered on large enough length scales.

There is a fixed point at $(x^*, y^*) = (0, 0)$. This is at zero temperature and infinite field and corresponds to a fully aligned configuration. Finally, and of greatest interest, the ferromagnetic fixed point occurs at $(x^*, y^*) = (0, 1)$ or zero temperature and zero field. Here the system is critical, correlations become long-ranged, and the scaling behaviour can be investigated.

The trajectories shown in Fig. 9.2 indicate how a system flows under the recursion eqns (9.12)–(9.14). Although they have been drawn as continuous, any one set of initial conditions will lead to a discontinuous trajectory. Note that the critical behaviour of the one-dimensional Ising model is governed by the least stable fixed point. Two relevant variables, temperature and magnetic field, must take the correct values before the system becomes critical.

9.1.3   Fixed points and scaling

Linearizing the recursion eqns (9.12)–(9.14) about the ferromagnetic fixed point $(x^*, y^*) = (0, 1)$ gives

$$x' \sim 4x, \qquad \epsilon' \sim 2\epsilon \qquad\qquad (9.15)$$

where $\epsilon = y - y^* \equiv y - 1$. Hence the eigenvalues of the linearized transformation are

$$\lambda_1 = 4, \qquad \lambda_2 = 2. \qquad\qquad (9.16)$$

Recognizing that we are dealing with a scale change by a factor $b = 2$ the corresponding exponents, defined by eqn (8.19), are

$$y_1 = 2, \qquad y_2 = 1. \qquad\qquad (9.17)$$

Hence, using eqn (8.26), the singular part of the free energy per spin near the fixed point can be written in the form

$$\bar{f}_s(x, \epsilon) = b^{-1} \bar{f}_s(b^2 x, b\epsilon). \qquad\qquad (9.18)$$

By iterating the transformation an arbitrary number of times b can be taken to have any value. (Strictly b can only be some power of 2 but an analytic continuation in b is possible and this turns out not to matter.) Therefore we choose $b^2 x = 1$ giving

$$\bar{f}_s(x, \epsilon) = \sqrt{x}\bar{f}_s(1, \epsilon/\sqrt{x}) = \sqrt{x}\tilde{f}_s(\epsilon/\sqrt{x}). \qquad (9.19)$$

Let us check that this form is correct. Using the transfer matrix method the free energy of the one-dimensional Ising model in a field was derived in Chapter 5. Translating eqn (5.32) into the variables used in eqn (9.19) and expanding for small ϵ and x one obtains

$$\bar{f}(x, \epsilon) = -K - \sqrt{x}\sqrt{1 + \epsilon^2/x} \qquad (9.20)$$

the singular part of which agrees with 9.19 if the scaling function is identified as

$$\tilde{f}(z) = \sqrt{1 + z^2} \qquad (9.21)$$

Note that the renormalization group has predicted that the free energy exhibits a scaling form near the critical point but has not given an explicit form for the scaling function. Derivatives of the free energy, such as the magnetization and the susceptibility, scale in a similar way near criticality (see problem 8.4).

Having obtained scaling forms for the thermodynamic functions it should be possible to find explicit values for the critical exponents. Because the transition takes place at zero temperature the usual definition of the exponents breaks down. However, a consistent set of definitions is obtained if the reduced temperature $t = (T - T_c)/T_c$ is replaced by $x = e^{-4K}$. The exponents $\alpha, \beta \ldots$ are then related to y_1 and y_2, given in eqns (9.17), by eqns (8.32)–(8.35) as usual.

9.1.4   The free energy

Finally we shall demonstrate that it is possible to calculate the total free energy for all temperatures by following the recursion eqns (9.12)–(9.14) as they flow to the high temperature sink. We shall consider the zero-field case to avoid complicated arithmetic.

The free energy is related to the renormalization of the constant term in the Hamiltonian. To see this explicitly divide the free energy after l iterations into two parts: $\bar{f}_0(x_l)$, the reduced free energy per spin for the Hamiltonian with the constant term put equal to zero, and C_l, the constant term itself. The recursion equation for the free energy, eqn (8.4), then becomes, iterating l times,

$$\bar{f}_0(x_0) - C_0 = 2^{-l}(\bar{f}_0(x_l) - C_l). \qquad (9.22)$$

The constant term renormalizes according to eqn (9.10), which may be rewritten

$$C_l = 2C_{l-1} + R_0(x_{l-1}).$$ (9.23)

By expressing C_{l-1} in terms of C_{l-2} and so forth eqn (9.23) becomes

$$2^{-l}C_l = \sum_{k=0}^{\infty} 2^{-(k+1)} R_0(x_k).$$ (9.24)

Combining eqns (9.22) and (9.24) gives an expression for the free energy at a coupling constant x_0,

$$\bar{f}_0(x_0) - C_0 = 2^{-l}\bar{f}_0(x_l) - \sum_{k=0}^{\infty} 2^{-(k+1)} R_0(x_k).$$ (9.25)

Although written here in the notation we have used for the one-dimensional Ising model, the formula (9.25) is a general result.

For large enough l the coupling constants in the first term on the right-hand side of eqn (9.25) tend to their value at a trivial fixed point where the free energy can usually be calculated. The second term can be obtained by numerically performing a sum over suitably weighted values of R_0 evaluated at each point on the trajectory (see problem 9.5).

For the special case of the one-dimensional Ising model, however, exact results can be obtained directly from eqn (9.22) and this is the approach followed here. As $l \to \infty$ the system approaches the attracting fixed point at $x^* = 1$. Here it behaves as a paramagnet with free energy per spin

$$\bar{f}_0(x_l) - C_l = -(\ln 2 + C_l) = \ln(\omega_l^{\frac{1}{4}}/2).$$ (9.26)

It follows immediately from eqns (9.22) and (9.26) that

$$\bar{f}_0(x_0) - C_0 = \lim_{l \to \infty} 2^{-l} \ln(\omega_l^{\frac{1}{4}}/2).$$ (9.27)

ω_l is easily calculated by a further change of variables. Rewriting eqns (9.12) and (9.13) (with $y = 1$) in terms of

$$u = (\omega x)^{\frac{1}{4}}/2, \qquad v = \tanh K = \frac{1 - \sqrt{x}}{1 + \sqrt{x}}$$ (9.28)

gives

$$u' = \frac{u^2(1+v)^2}{1+v^2}, \qquad v' = v^2. \tag{9.29}$$

As $l \to \infty$, $x_l \to 1$. Hence eqn (9.27) becomes

$$\bar{f}_0(x_0) - C_0 = \lim_{l\to\infty} 2^{-l}\ln u_l. \tag{9.30}$$

Iterating eqns (9.29) a few times it becomes apparent that the general expression for u_l is

$$\ln u_l = 2^l \ln u_0 + 2^l \ln(1+v_0) - \ln(1+v^{2^l}). \tag{9.31}$$

As v is less than unity for all non-zero temperatures the last term in this expression drops out in the limit $l \to \infty$. Hence substituting eqn (9.31) into eqn (9.30) gives

$$\bar{f}_0(x_0) - C_0 = \ln u_0(1+v_0) = -C_0 - \ln 2\cosh K \tag{9.32}$$

where we have translated into more familiar variables to regain the expression (5.32) for the free energy obtained from the transfer matrix method in Chapter 5.

9.2 Higher dimensions

We now turn to consider real-space renormalization group transformations in higher dimensions. One might expect a richer fixed-point structure because the critical point is no longer at zero temperature. This is indeed the case and it will enable us to demonstrate more easily than in one dimension the importance of relevant and irrelevant variables and universality. The drawback is the necessity for approximations: only very rarely is it possible to write down exact recursion equations in two and three dimensions.

Consider a decimation transformation for the two-dimensional Ising model. This is one of the simplest of a host of approximation methods presented in the literature, but it provides a good illustration of the main points. At the end of this section other approaches to the problem of finding useful approximate real-space recursion relations will be mentioned briefly but many of these, ingenious and complicated as they can be, have their natural home in the review articles listed at the end of this book.

We shall work with the square lattice shown in Fig. 8.1. Decimating out, or doing the partial trace, over all spins marked with a cross gives a new square lattice related to the first by a scale change $b = \sqrt{2}$. For clarity the spins on the original lattice that do not survive the transformation are labelled s and those that do are labelled t. The partition function can be written as a product of terms, each of which contains only one s-spin, for example

$$\exp K s_{00}(t_{01} + t_{0-1} + t_{10} + t_{-10}) \tag{9.33}$$

where our notation is made explicit by Fig. 9.3. Taking the trace over s_{00} gives

$$2 \cosh K(t_{01} + t_{0-1} + t_{10} + t_{-10}) \tag{9.34}$$

which can be rewritten as

$$\exp\{a(K) + b(K)(t_{-10}t_{01} + t_{01}t_{10} + t_{10}t_{0-1}$$
$$+ t_{0-1}t_{-10} + t_{-10}t_{10} + t_{0-1}t_{01}) + c(K)t_{-10}t_{01}t_{10}t_{0-1}\} \tag{9.35}$$

if

$$a(K) = \ln 2 + (\ln \cosh 4K + 4 \ln \cosh 2K)/8, \tag{9.36}$$
$$b(K) = (\ln \cosh 4K)/8, \tag{9.37}$$
$$c(K) = (\ln \cosh 4K - 4 \ln \cosh 2K)/8. \tag{9.38}$$

To show that this is correct one has to check that expressions (9.34) and (9.35) are indeed equal for all sixteen possible values of the t variables.

The interactions generated on the renormalized lattice by taking the trace over s_{00} are not just between nearest neighbours—there are also second neighbour and four-spin terms. By considering similar contributions from tracing over all the s-spins it is apparent that the renormalized Hamiltonian is (ignoring the constant term)

$$\bar{\mathcal{H}}' = 2b(K) \sum_{\langle ij \rangle} t_i t_j + b(K) \sum_{[ij]} t_i t_j + c(K) \sum_{sq} t_i t_j t_k t_l \tag{9.39}$$

where $[ij]$ denotes a sum over second neighbours and sq a sum over four neighbours around an elementary square on the renormalized lattice. The factor of 2 arises in the nearest neighbour term because of contributions from each of the adjacent s-spins.

So, having started with just nearest neighbour interactions, a Hamiltonian with second neighbour and four-spin couplings has been generated. Further iterations of the renormalization group will generate

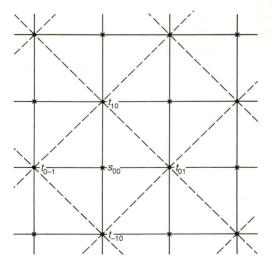

Fig. 9.3. Decimation of the square lattice by a scale factor $b = \sqrt{2}$. Spins that survive the transformation are labelled t; those that do not are labelled s.

long-range and multi-spin interactions of arbitrary complexity. More-over, even a second iteration cannot be performed exactly because the spins to be decimated out appear in complicated combinations in the Hamiltonian (9.39) rendering a simple decomposition into terms such as (9.33) impossible.

   An approximation is needed which will truncate the parameter space after each iteration so that it remains closed; that is no new interactions are generated at each step. The hope is, and this is borne out by numerical examples, that the short-range interactions dominate the critical behaviour. For illustrative purposes we shall follow Wilson[1] in considering a lattice with first and second neighbour couplings K and L and ignore any higher-order terms generated by the renormal-ization group. Moreover we shall keep only terms of order K^2 and L. This considerably simplifies the arithmetic without changing the physics. One obtains the recursion equations

$$K' = 2K^2 + L, \qquad L' = K^2. \tag{9.40}$$

The terms of order K^2 follow from renormalizing the first neighbour interaction and the expansion of $b(K)$ in eqn (9.39). The term L follows

[1]Wilson, K. G. (1975). *Reviews of Modern Physics*, **47**, 773.

from bonds like $t_{10}t_{01}$ which correspond to second neighbours on the original lattice but first neighbours on the renormalized lattice.

The recursion eqns (9.40) have a ferromagnetic fixed point at $(K^*, L^*) = (\infty, \infty)$, a paramagnetic fixed point at $(K^*, L^*) = (0, 0)$ and a non-trivial fixed point at $(K^*, L^*) = (1/3, 1/9)$. Linearizing about the non-trivial fixed point gives one relevant and one irrelevant eigenvalue, $\lambda_1 = 1.721$ and $\lambda_2 = -0.387$ respectively. Using eqns (8.19) and (8.40) the critical exponent ν is directly related to λ_1

$$\nu = \frac{\ln b}{\ln \lambda_1} = 0.638 \qquad (9.41)$$

recalling that $b = \sqrt{2}$ for the scaling we are considering. The exact value is 1.

The basin of attraction of the non-trivial fixed point is a line as expected if there is one irrelevant variable. This approximates the line of critical values of the two-dimensional Ising model with first and second neighbour interactions $K_c(L)$. The point $K_c(0) = 0.392$ flows into the fixed point and provides the estimate of the transition temperature of the Ising model with nearest neighbour interactions only. The exact value is 0.441.

The most important prediction of this simple approximation is that the critical surface, $K_c(L)$, is governed by a single fixed point. Hence the critical exponents are independent of the relative values of the first and second neighbour interactions at criticality. This is an explicit example of how the property of universality follows from the renormalization group formalism.

If the parameter space is large enough to allow the correct topology the fixed point and flow structure are often surprisingly insensitive to the exact form of the approximate recursion relations. Even very crude renormalization groups can sometimes accurately demonstrate the important topologies within parameter space.

Obtaining accurate quantitative results is much more difficult. The numerical values calculated above for the two-dimensional Ising model were reasonable given the severity of the approximation used. Including more interactions in the recursion equations very quickly becomes prohibitively complicated. It often allows a better estimate of the critical temperature and the exponents, but there is no controlled way of deciding which interactions it is important to include. It is far from true that enlarging the parameter space automatically implies more accurate numerical results.

Moreover, any decimation transformation has a more serious defect built in. Because the magnitude of the spin remains unchanged by the transformation, the spin rescaling factor, $c(b)$, defined by eqn (8.43), is of necessity equal to unity. This incorrectly forces η to be zero in two dimensions and implies that the very existence of a fixed point distribution is an artefact of the approximate nature of the recursion equations[2].

Different approaches to the construction of real-space recursion equations are legion. In block-spin transformations, such as that described in Section 1.2.1, clusters of spins are replaced by a single spin which takes the same value as the majority of spins in the original cluster. In variational renormalization groups parameters are introduced into the recursion equations and tuned to reproduce known results correctly or to extremize the free energy. The fixed point topology in parameter space is often insensitive to the particular approximation used. Accurate critical exponents can be obtained too (particularly if the exact answer is there for comparison!) but, because of the uncontrolled nature of the approximations, it is difficult to know how trustworthy the results are in new applications.

Rather than dwell at length on the details of how to construct real-space renormalization groups I shall turn the discussion in two disparate directions. The next topic is a description of renormalization for the q-state Potts model. This is included as an illustration of the importance of choosing the right parameter space, of a more complicated fixed-point structure, and of how new results followed immediately from considering that structure. The second, the Monte Carlo renormalization group, is a marrying of the ideas of Monte Carlo simulations and the renormalization group which does allow accurate and controlled determinations of critical temperatures and critical exponents.

9.3   The q-state Potts model

We cite the q-state Potts model as an example where understanding the correct topology of the flows within parameter space has led to useful new results. Recall that the q-state Potts model is described by the Hamiltonian

$$\mathcal{H} = -J \sum_{\langle ij \rangle} \delta_{s_i s_j}, \qquad s_i = 1, 2, \ldots q. \tag{9.42}$$

It was known before the advent of the renormalization group that the transition to the ferromagnetic state is continuous for $q \le q_c$ but first-order for $q > q_c$ where $q_c = 4$. However, early renormalization group

---

[2]Wilson, K. G. (1975). *Reviews of Modern Physics*, **47**, 773.

calculations were unable to reproduce this crossover.

In the late 1970s a group of scientists in the USA realized that the problem arose because the parameter space considered did not include the right variables to allow the correct fixed-point topology[3]. In place of the Hamiltonian (9.42) they used as their starting point the dilute q-state Potts model

$$\mathcal{H} = \sum_{\langle ij \rangle} (-L - J\delta_{s_i s_j}) t_i t_j + \Delta \sum_i t_i,$$

$$s_i = 1, 2 \ldots q; \quad t_i = 0, 1. \tag{9.43}$$

This reduces to the usual Potts model if all the $t_i = 1$ (which will be the situation if the chemical potential $\Delta = -\infty$) but allowing $t_i = 0$ gives the possibility of vacant sites in equilibrium with the spins in the system.

The physical motivation behind this idea was that to map a region where the spins took many of the q possible values on to any one of those values would seriously overestimate the order in the system. Such areas were better represented by a vacancy on the renormalized lattice.

The flows of the Hamiltonian (9.43) under a real-space renormalization transformation are shown schematically in Fig. 9.4. The value of q is unchanged by the transformation so all flows are along cross-sections of constant q. The variable L which appears in the Hamiltonian is necessary to close the parameter space but is not pertinent to our argument and is suppressed from the diagram.

In Fig. 9.4 the surface ABCD separates the ferromagnetic and the disordered phases. Within this surface there is a line of critical fixed points, EF, where relevant variables only flow out of the critical surface and a line of tricritical fixed points, GF, where there is an extra relevant variable within the critical surface itself. The two lines meet continuously at F. CD is a line of fixed points which describe first-order phase transitions. The intersection of the critical surface ABCD with the plane $e^{\Delta - 3J} = 0$ defines the critical temperature $kT_c(q)/J$ of the usual pure Potts model (eqn 9.42). One sees immediately that, as the renormalization group transformation is iterated, flows that begin here are attracted by a critical fixed point for $q < q_c$ and by a first-order fixed point for $q > q_c$. Without the introduction of vacancies there would not have been a large enough parameter space for such a topology to exist.

More can be inferred from Fig. 9.4. Moving along the line of critical fixed points EF, the critical exponents will change continuously and

---

[3]Nienhuis, B., Berker, A. N., Riedel, E. K., and Schick, M. (1979). *Physical Review Letters*, **43**, 737.

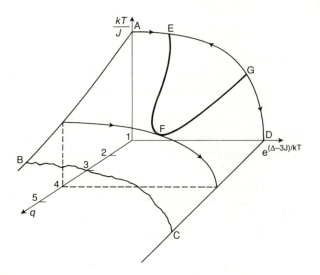

Fig. 9.4. Flows within parameter space for a real-space renormalization group transformation of the q-state Potts model with vacancies. ABCD is the surface separating the ferromagnetic and disordered phases. EF and GF are lines of critical and tricritical fixed points respectively. CD is a line of first-order fixed points. After Nienhuis, B., Berker, A. N., Riedel, E. K., and Schick, M. (1979). *Physical Review Letters*, **43**, 737.

hence it should be possible to write a given exponent as a function of q. Such a form had been conjectured prior to the calculation described above. But we have seen that the tricritical points lie on a different branch of the same curve. Hence it should be possible to extend the conjectured formula analytically to encompass the variation of the tricritical exponents with q. This is indeed the case and the result has now been proved exactly.

9.4   The Monte Carlo renormalization group

We describe how a combination of the Monte Carlo method described in Chapter 7 and the ideas of the real-space renormalization group can lead to very precise estimates of critical parameters. It is apparent that the major drawback of real-space renormalization group approaches is the necessity for severe and uncontrolled truncations of the parameter space because of the computational difficulties inherent in including very many interactions. A problem associated with the Monte Carlo method is the difficulty of calculating critical properties because, as the correlation length becomes infinite, finite-size corrections obscure the critical singularities. The concatenation of both approaches in the Monte Carlo renormalization group, first described by Ma in 1976[4], provides a practical way of overcoming both difficulties.

A Monte Carlo renormalization group transformation is performed by using the standard Monte Carlo procedure to generate a large number of uncorrelated equilibrium states for a model near criticality. Each configuration is then renormalized by means of a block spin transformation like that described in Section 1.2.1. The system is divided into blocks, typically of linear dimension two spins, and each block is replaced by a spin which has the same value as the majority of the original spins. Any 'ties' can be broken by means of a random number generator. This procedure is easily carried out numerically.

The renormalization transformation is repeated as often as is feasible: the lattice must always remain large enough to include all the interaction parameters considered. Because the original Hamiltonian is chosen to be as close to criticality as possible, repeated iterations move it towards the fixed point. The matrix **A**, defined by eqn (8.17), can be calculated from correlation functions between spin operators on the old and new lattices at each step of the transformation. Hence a sequence of estimates of the critical exponents which converges towards the fixed point values results.

---

[4]Ma, S.-K. (1976), *Physical Review Letters*, **37**, 461.

Practical calculations can include 20 interactions in the analysis without any numerical difficulties. Extra interactions can be added to provide a check on the consistency of the method. Results for the three-dimensional Ising model[5], $K_c = 0.221654(6)$, $\nu = 0.629(4)$, and $\eta = 0.031(5)$, are consistent with, and achieve the same precision as, series calculations.

9.5 The ϵ-expansion

An alternative approach to the construction of renormalization group recursion equations is the ϵ-expansion. This relies heavily on the techniques of field theory so only a qualitative description is of order here.

The starting point is a Hamiltonian, similar in form to eqn (4.35), but with the addition of a term quartic in the magnetization

$$\mathcal{H} \sim \int_q (\tilde{r}+q^2) \mid m(q) \mid^2 + \int_{q_1} \int_{q_2} \int_{q_3} \tilde{u}\, m(q_1)m(q_2)m(q_3)m(-q_1-q_2-q_3).$$

$$(9.44)$$

This Hamiltonian provides an approximation to a continuum version of the Ising model and is expected, because of universality, to share the same critical exponents.

The renormalization group equations are obtained[6] by integrating over large momenta rather than, as is the case for the real-space approaches we have considered up to now, short length scales. To first order in u

$$r' = 4\{r + 3cu/(1+r)\}, \qquad (9.45)$$

$$u' = 2^\epsilon \{u - 9cu^2/(1+r)^2\} \qquad (9.46)$$

where r and u are proportional to $\tilde{r}$ and $\tilde{u}$. The parameter $\epsilon = 4 - d$ depends on the dimension. u is of order ϵ and hence an expansion in u can also be considered as an expansion around four dimensions.

For $d > 4$ ($\epsilon < 0$) the stable fixed point lies at $r^* = u^* = 0$. This fixed point has mean-field exponents, independent of the dimensionality. However, for $d < 4$, u becomes an extra relevant variable at the mean-field or Gaussian fixed point and drives the system away from it. The critical behaviour is now governed by a fixed point at $r^* = -(4\epsilon \ln 2)/9$, $u^* = (\epsilon \ln 2)/9c$ where the exponents take Ising values which do depend on the dimensionality. Thus the ϵ-expansion has

---

[5]Pawley, G. S., Swendsen, R. H., Wallace, D. J., and Wilson, K. G. (1984). *Physical Review*, **B29**, 4030.

[6]Wilson, K. G. and Kogut, J. (1974). *Physics Reports*, **12**, 75.

predicted the existence of an upper critical dimensionality at $\epsilon = 0$ or $d = 4$ above which critical exponents take mean-field values. This confirms the heuristic argument given in Section 4.5.

It is of interest to ask how useful the ϵ-expansion is for the quantitative prediction of three-dimensional critical exponents. Putting $\epsilon = 1$ gives $\gamma = 1.17$, 1.245, and 1.195 to order ϵ, ϵ^2, and ϵ^3 respectively. Calculation to higher orders in ϵ is usually prohibitively difficult. The accepted value from series and numerical work is 1.24 which compares well with the value obtained to order ϵ^2. This appears to be generally true for reasons which are not fully understood.

9.6   Problems

9.1 (i) Show that the recursion relations for a decimation transformation by a scale factor b for a one-dimensional spin model may be written in terms of the transfer matrix $\mathbf{T}$ for that model as

$$\mathbf{T}(\{K'\}) = \mathbf{T}^b(\{K\})$$

where $\{K\}$ and $\{K'\}$ are the coupling constants on the original and renormalized lattices respectively.
(ii) Check that for the spin-1/2 Ising model this approach correctly reproduces the recursion relations (9.8)–(9.10) for $b = 2$.

9.2 A simple, approximate way of changing the length scale of a lattice is the Migdal–Kadanoff transformation[7]. This is illustrated for the square lattice in Fig. 9.6. It comprises two steps:

1. Half the bonds are moved to change the scale of the lattice by a factor $b = 2$. This is compensated for by doubling the strength of the remaining bonds.

2. The sites marked by a cross are removed by a one-dimensional decimation transformation.

(i) Show that the Migdal–Kadanoff recursion equation for the spin-1/2 Ising model on a square lattice is

$$x' = 2x^2/(1 + x^4)$$

where $x' = e^{-2K'}$ and $x = e^{-2K}$.
(ii) Find the fixed points of the transformation and compare the

[7]Kadanoff, L. P. (1976). *Annals of Physics*, **100**, 359.

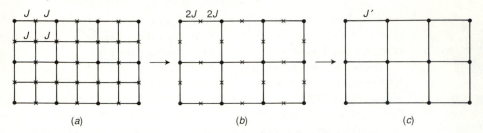

Fig. 9.5. The steps in a Migdal–Kadanoff renormalization group transformation for a square lattice. (a) The original lattice with coupling constant J. (b) Half the bonds are removed to change the scale of the lattice by a factor $b = 2$. To compensate for this the strength of the remaining bonds is doubled. (c) The sites marked with a cross are removed using a one-dimensional decimation transformation to give a renormalized coupling constant J'.

Migdal–Kadanoff result for the critical temperature to the exact value.

(iii) By linearizing around the non-trivial fixed point find the Migdal–Kadanoff approximation to the correlation length exponent ν.

[Answers: $x_c = 0.544$, $\nu = 1.34$.]

9.3 (i) Use the same approach as in problem 9.1 to find recursion relations for the spin-1 Ising model

$$\mathcal{H} = -J \sum_{\langle ij \rangle} s_i s_j - \sum_i C, \quad s_i = \pm 1, 0.$$

The recursion equations are inconsistent showing that new interactions must be generated by the decimation procedure.

(ii) Show that if all possible even interactions are included in the original Hamiltonian

$$\mathcal{H} = -J \sum_{\langle ij \rangle} s_i s_j - K \sum_{\langle ij \rangle} s_i^2 s_j^2 - D \sum_i s_i^2 - \sum_i C, \quad s_i = \pm 1, 0.$$

a set of consistent recursion equations can be generated, namely,

$$x' = \frac{(1 + y + z)x}{1 + x^2 + y^2},$$

$$y' = \frac{2y + x^2}{1 + x^2 + y^2},$$

$$z' = \frac{z^2 + 2x^2}{1 + x^2 + y^2}$$

where

$$x = e^{-\beta(J + K + D/2)}, \quad y = e^{-2\beta J}, \quad z = e^{-\beta(J + K + D)}.$$

9.4 Using a Migdal–Kadanoff approximation such as that described in problem 9.2 but with the one-dimensional decimation preceding the bond moving the recursion equations for the spin-1 Ising model derived in problem 9.3 can be turned into those appropriate to a two-dimensional system. Writing $(J', K', D') \Rightarrow (2J', 2K', 2D')$

$$x' = \left\{ \frac{(1 + y + z)x}{1 + x^2 + y^2} \right\}^2,$$

$$y' = \left\{ \frac{2y + x^2}{1 + x^2 + y^2} \right\}^2,$$

$$z' = \left\{ \frac{z^2 + 2x^2}{1 + x^2 + y^2} \right\}^2.$$

(i) Show that $x = 0$ is an invariant subspace of the equations, that is, flows which start there remain there. The recursion relations restricted to this subspace are those used as an example in Section 8.4.

(ii) Show that there is a fixed point at

$$(x, y, z) = (0.326, 0.015, 0.922)$$

with two relevant and one irrelevant eigenvalue. This is a tricritical fixed point which governs a line of tricritical points in the phase diagram.

(iii) Show that there is a fixed point at

$$(x, y, z) = (1/4, 1/4, 1)$$

with three relevant eigenvalues corresponding to the values of the interaction parameters at which the Hamiltonian has the symmetry of the 3-state Potts model.

9.5 The free energy of the one-dimensional Ising model can be calculated using the renormalization group through eqn (9.25). For

the one-dimensional spin-1/2 Ising model
(i) find an expression for $R_0(x_k)$.
(ii) explain why $2^{-l}\bar{f}_0(x_l) \to 0$ as $l \to \infty$.
(iii) calculate the trajectory sum in eqn (9.25) numerically. Hence find values for $\bar{f}_0(x_0)$ for chosen values of x_0. Check your answers against the exact result (9.32).

9.6 The recursion equations obtained by performing an ϵ-expansion on the continuum version of the Ising model are, to first-order in u,

$$r' = 4\{r + 3cu/(1+r)\}, \qquad\qquad (9.47)$$
$$u' = 2^\epsilon\{u - 9cu^2/(1+r)^2\} \qquad\qquad (9.48)$$

where $\epsilon = 4 - d$.
(i) Show that there are fixed points at $(r^*, u^*) = (0,0)$ and $(r^*, u^*) = (-\{4\epsilon \ln 2\}/9, \{\epsilon \ln 2\}/9c)$ with eigenvalues $\lambda_1 = 4$, $\lambda_2 = 1 + \epsilon \ln 2$ and $\lambda_1 = 4 - \{4\epsilon \ln 2\}/3$, $\lambda_2 = 1 - \epsilon \ln 2$ respectively.
(ii) Show that, for $d > 4$, the Gaussian fixed point at $(r^*, u^*) = (0,0)$ has one relevant and one irrelevant eigenvalue and the Ising fixed point has two relevant eigenvalues but that, for $d < 4$, the situation is reversed. This means that the behaviour on the one-dimensional critical surface is controlled by the Gaussian fixed point for $d > 4$ and the Ising fixed point for $d < 4$.
(iii) Find the exponent ν for the Gaussian fixed point and show that it is independent of the dimensionality and equal to the mean-field value. This is the fixed point that describes mean-field behaviour.
(iv) Show that, for the Ising fixed point, $\nu = 1/2 + \epsilon/12$ to first order in ϵ.

Further reading

A selection of references is listed below for those wishing to pursue the subject further[8]

1. Early, clear accounts:

   (a) Stanley, H. E. (1971). *Introduction to phase transitions and critical phenomena.* (Oxford University Press, Oxford).

   (b) Fisher, M. E. (1967). *Reports on Progress in Physics*, **30**, 615.

2. An expanding series of volumes containing review articles which provides a standard work of reference to researchers in the field:

   (a) Domb, C. and Green, M. S. (eds) (1972–76). *Phase transitions and critical phenomena*, Vols 1–6. (Academic Press, London).
   Domb, C. and Lebowitz, J. L. (eds) (1983–). *Phase transitions and critical phenomena*, Vols 7–. (Academic Press, London).

3. Books dealing with related topics:

   (a) Thompson, C. J. (1988). *Classical equilibrium statistical mechanics.* (Clarendon Press, Oxford).

   (b) Patashinskii, A. Z. and Pokrovskii, V. I. (1979). *Fluctuation theory of phase transitions.* (Pergamon, Oxford).

---

[8]See also Table 1.1 for references to books describing different types of phase transition.

(c) Binney, J. J., Dowrick, N. J., Fisher, A. J., and Newman, M. E. J. (1992). *The modern theory of critical phenomena.* (Clarendon Press, Oxford).

(d) Baxter, R. J. (1982). *Exactly solved models in statistical mechanics.* (Academic Press, London and San Diego).

(e) Amit, D. J. (1984). *Field theory, the renormalization group, and critical phenomena,* (2nd edn). (World Scientific, Singapore).

(f) Parisi, G. (1988). *Statistical field theory.* (Addison-Wesley, Wokingham).

(g) Huang, K. (1987). *Statistical mechanics,* (2nd edn), chs 14–18. (Wiley, New York).

4. Books and review articles describing the renormalization group:

(a) Wilson, K. G. and Kogut, J. (1974). *Physics Reports*, **12**, 75.

(b) Fisher, M. E. (1974). *Reviews of Modern Physics*, **46**, 597.

(c) Wilson, K. G. (1975). *Reviews of Modern Physics*, **47**, 773.

(d) Barber, M. N. (1977). *Physics Reports*, **29**, 1.

(e) Wallace, D. J. and Zia, R. K. P. (1978). *Reports on Progress in Physics*, **41**, 1.

(f) Wilson, K. G. (1983). *Reviews of Modern Physics*, **55**, 583.

(g) Pfeuty, P. and Toulouse, G. (1977). *Introduction to the renormalization group and to critical phenomena.* (Wiley, London).

(h) Ma S.-K. (1976). *Modern Theory of Critical Phenomena.* (Addison-Wesley, Wokingham).

5. Numerical methods:

(a) Binder, K. (ed.) (1986). *Monte Carlo methods in statistical physics,* (2nd edn). (Springer-Verlag, New York).

(b) Binder, K. (ed.) (1987). *Applications of the Monte Carlo method in statistical physics,* (2nd edn). (Springer-Verlag, New York).

(c) Allen, M. P. and Tildesley, D. J. (1987). *Computer simulations of liquids.* (Clarendon Press, Oxford).

Index